Environmental, Health and Humanity Issues in the Down Danubian Region

Multidisciplinary Approaches

Environmental, Health and Humanity Issues in the Down Danubian Region

Multidisciplinary Approaches

Proceedings of the 9th International Symposium on Interdisciplinary Regional Research

21-22 June 2007 University of Novi Sad

Editors

Dragutin Mihailović

Mirjana Vojinović Miloradov

University of Novi Sad, Serbia

World Scientific

NEW JERSEY • LONDON • SINGAPORE • BEIJING • SHANGHAI • HONG KONG • TAIPEI • CHENNAI

Published by

World Scientific Publishing Co. Pte. Ltd.
5 Toh Tuck Link, Singapore 596224
USA office: 27 Warren Street, Suite 401-402, Hackensack, NJ 07601
UK office: 57 Shelton Street, Covent Garden, London WC2H 9HE

British Library Cataloguing-in-Publication Data
A catalogue record for this book is available from the British Library.

ENVIRONMENTAL, HEALTH AND HUMANITY ISSUES IN THE DOWN DANUBIAN REGION
Multidisciplinary Approaches

ISBN-13 978-981-283-439-3
ISBN-10 981-283-439-7

Typeset by Stallion Press
Email: enquiries@stallionpress.com

Printed in Singapore by World Scientific Printers

CONTENTS

PREFACE

We live in a changing world. This thought expresses a historical fact entirely present in man's consciousness regardless of whether we consider him as an individual or a social human being. The awareness about the changing world always introduces an amount of anxiety into man's life that can be defined by the following question: *Can man preserve the existing world for the future of his children?*

To preserve this world for the future of our children, we must all strive for sustainable development — a development that meets the needs of the present generation without compromising options and resources the future generations will use to meet their own needs. It implies environmentally sound development in societies and regions free from threats to life and property. Human security is an essential ingredient of sustainability, which is increasingly threatened by extreme events, both natural and human-induced.

Regardless of the word *balance* being used either globally or locally in any given context, it is undoubtedly the keyword in the increasing number of environmental problems. The underlined sketch is a proper introduction to the question: *Why are the environmental problems in focus now?* One particular answer can be found in the hierarchy of the main scientific problems for the 21st century as seen by the scientific community. According to them, in the the 21st century the world of science will be occupied by the problems linked mainly to superconductivity, quantum teleology, extra-terrestrial contacts and environmental problems primarily expressed through the climate change problems. A unique characteristic of those problems is the question of different aspects of the existence of an individual human being. Those questions are: technological capability, origin of the consciousness and survival on the Earth. This is the first time in the history of science that the environmental problems take place at the frontline of science. The question of *why is this happening now and why it will go on happening in the future* could be answered by the well known fact that in the scientific as well as other worlds the main "drama of the event" takes place at the interface between either two media or two states.

Damages in lives lost and destroyed property are increasing in frequency and magnitude. Increased pollution in the atmosphere is causing gradual increase of the temperature. The slowly rising sea level is threatening low lying countries with more severe storm surges and some scientists suspect that recent large floods, heat waves and drought in Central Europe and the lower Danubian region are indicators of more severe hazardous events in the future. Even more dramatic are human-induced environmental changes caused by population growth, uncontrolled urbanisation and development, and regional conflicts. They result in changes of social and cultural structures, public health, changes of land use and related phenomena, and generating environmental refugees. The environment becomes increasingly endangered, especially when man encroaches on and modifies fragile environments.

The 9th International Symposium on Interdisciplinary Regional Research — ISIRR 2007 took place on 21–22 June 2007 at the University of Novi Sad, Serbia. This Symposium is the result of tripartite regional cooperation between the University of Novi Sad, Regional Committee of the Hungarian Academy of Sciences — Szeged (Hungary) branch and Interdisciplinary Committee for Research and Development of the Romanian West Zone — Timisoara (Romania), a cooperation that has been ongoing since 1996. The results of joint research projects are presented at the international symposia "Interdisciplinary Regional Research — ISIRR", which are being held cyclically in one of the three countries.

The University of Novi Sad hosted the 3rd and 6th ISIRR Symposiums in 1998 and 2002 respectively and accepted the organisation of the 9th Symposium by the decision of its Teaching and Research Council. Having in mind the interdisciplinary character of the Symposium, the Organising Committee was formed from the professors in charge of University Centres within the Association of Centres for Interdisciplinary and Multidisciplinary Studies and

Research (ACIMSI), University of Novi Sad, Serbia. Having in mind its interdisciplinary character, the Symposium was carried out through the six sections with presentations in the form of plenary lectures as well as oral and poster presentations.

This book comprises materials selected among all the submitted papers. It contains four chapters: The Medical Issues (Chapter One), The Agricultural and Food Technology Issues (Chapter Two), The Ecological Issues (Chapter Three) and Humanity Issues (Chapter Four).

Part One deals with the medical problems in the lower Danubian region. Keynote papers consider medical imaging in general and Positron Emission Tomography (PET). The following chapter papers are devoted to various problems related to the application of DICOM/JPEG2000 Client/Server Architecture, radionuclide application in therapy and diagnostics and densitometry measurements. Other papers consider problems related to traveller's thrombosis, acute limb ischemia and dentures technology.

Part Two considers the agricultural, technological and developmental open horizons in natural resources of the lower Danubian region. Through applied and fundamental disciplines ranging from biochemistry to physiology, microbiology, applied genetics and food technology, the authors searched for innovations that promote sustainable management of crops and other natural resources in this agriculturally developed region of Europe. The first paper is devoted to the plant antioxidant activity of wild and cultivated *Allium* species, while the second one deals with the antioxidant systems in some rare and endangered species from the southern part of the Pannonian plain. The group of biochemistry papers is closed with the studies of (i) distribution of different forms of cadmium in the intercellular space and cadmium content in roots, stems and leaves of young sugar beet plants and (ii) effect of nickel, cadmium and molybdenum on sugar beet nitrate and protein metabolisms. The first paper that comes from the physiological background deals with the effect of short-term water deficiency on vascular tissue of petiole and leaf lamina of sugar beet, while the other three treat the effect of nitrate concentrations on physiology of different leaf age groups of poplar. Following is the paper in which the authors assess the effects of the application of Nitragin (microbiological fertilizer for soybean) and PGPR (Plant Growth Promoting Rhizobacteria) on soybean yield. The agricultural part of the chapter is ending with two papers from the field of the applied genetics studying cytological characteristics of regenerants obtained from another culture of wheat and tolerance to excess boron in wheat. The food technology part of Part Two comprises three papers. They consider problems related to the technology of plant and animal processing, i.e. (i) products from *Amaranthus* grain, (ii) application of medical plant extracts in biscuits and (iii) fatty acid composition of milk of the tsigai sheep, a variety specifically bred in the lower Danubian region.

Part Three includes papers dealing mostly with environmentally sound ecological modelling approaches, applied modelling examples and environmental experiments in the lower Danubian region. Four groups of topics can be identified among these papers. The first group contains topics of general theory of modelling for different applications, their challenges and problems. It comprises three papers giving several examples such as support to vector machines or nested ecological modelling and their applications. The second group of papers gives examples of applied modelling for climate change impact research on agriculture. These range from the application of spatial analogue techniques of climatically similar regions to impacts of climate change on selected crops as well as changed microclimatic conditions within crop stands and related stress levels. Finally, the last paper from this group is focused on the crop model calibration and validation as well as the application of a weather generator. The third group of papers highlights application of risk analysis and management methods applied to past and future climatic conditions as well as to certain crops and hazardous gases released in the pharmaceutical industry. Risk analysis, risk management and uncertainties under future climate scenarios are the most topical and increasingly important research topics in climate impact and adaptation research worldwide, because they play an important role in the transfer of this knowledge to decision makers. Two papers finalising this group of papers consider (i) microclimatic conditions and quantitative processes within current or future climate, such as the impact of different methods on uncertainties in lake water budget and (ii) modelling efforts in the pharmacokinetics system with time delay. The last two papers of Part Three, without direct modelling efforts, give examples of investigations, e.g. dealing with characterisation of gas/particle partitioning polychlorinated biphenyls (PCBs) and polycyclic aromatic hydrocarbons (PAHs) in the area of Kragujevac during the NATO operation in Serbia, and the importance of possible corn production in Hungary in the future, seen through the lens of the climate change.

In Part Four, the papers deal with several aspects of interdisciplinary relation of humanity issues and society through ecology, economy, language and literature. The very first paper analyses statements of interviewed citizens of the city of Novi Sad (Serbia) on measures of local environmental policy and stakeholders who participate in environmental policy actions. Following are two papers based on economic issues. They consider (i) job satisfaction in the state-owned sector of Serbia and (ii) the first public private partnership in Serbia whose socio-economic environment is fruitful for the implementation of new university-industry-government partnership. The last group of papers concern gender relations, Balkan culture and literature. They are a kind of pioneering work which addresses the struggle of getting gender studies recognised as a degree programme in the lower Danubian countries over the past two decades. The first paper from this group focuses on the role of male/female addresses on the same and opposite sexes and investigates how these addresses reflect today's Balkan culture, where women are traditionally in the background and men are in the foreground. The next paper deals with metaphors of cannibalism and the role they play in the female protagonists' reconsideration of their identities in several novels, while in the last one the author discusses two novels from a double perspective: the feminine condition and the role that art plays in the life of the characters.

Not all papers represent a new and original contribution, but they are interesting and give a scientific review and correct approach to the considered problems in a comprehensive and critical way. We are proud that young researchers took part with their first papers, giving valuable contribution to this Symposium.

This book is highly supported by the following people, who invested a large effort in its creation: Miss Maja Đogo, Miss Ana Pavlović and Miss Svetlana Vujović. Several state institutions financially supported the publication of this book. They are: Provincial Secretariat for Environmental Protection and Sustainable Development, Provincial Secretariat for Science and Technological Development, Provincial Secretariat for Education and Culture and Ministry of Science, Serbia. Dr. Dušan Sakulski from the United Nations University, Bonn (Germany) gave us his help, highly exceeding the level of common collegial support. Editors are deeply grateful to all of them.

Dragutin T. Mihailovic
Mirjana Miloradov

PART 1

MEDICAL ISSUES

CHAPTER 1

MEDICAL IMAGING — INDISPENSABLE MEDICAL TOOLS

SLOBODANKA STANKOVIĆ AND OLIVERA KLISURIĆ

Faculty of Sciences, University of Novi Sad, Novi Sad, Serbia

Physics and medicine are close allies. Ever since the discovery of X-rays and their potential for medical imaging, physicists have been vital to the advancement of medicine. Fundamental research in acoustics, electromagnetism and particle and nuclear physics has resulted with an array of indispensable medical tools. MRI (using microwaves), CAT scans (using X-rays), PET scans (using gamma rays), ultrasound scans (using sound waves) and various types of radiotherapy are among the physics-based devices that help doctors diagnose and treat ailments ranging from broken bones to cancer.

1. Introduction

Medical imaging refers to the techniques and processes used to create images of the human body (or parts thereof) for clinical purposes (medical procedures seeking to reveal, diagnose or examine disease) or medical science (including the study of normal anatomy and function).

To a human observer, the internal structures and functions of the human body are not generally visible. However, by various technologies, images can be created through which a medical professional can look into the body to diagnose abnormal conditions and guide therapeutic procedures. Medical image is a window to the body. No image window reveals everything. Different medical imaging methods reveal different characteristics of the human body. With each method, the image quality and structure visibility can vary considerably, depending

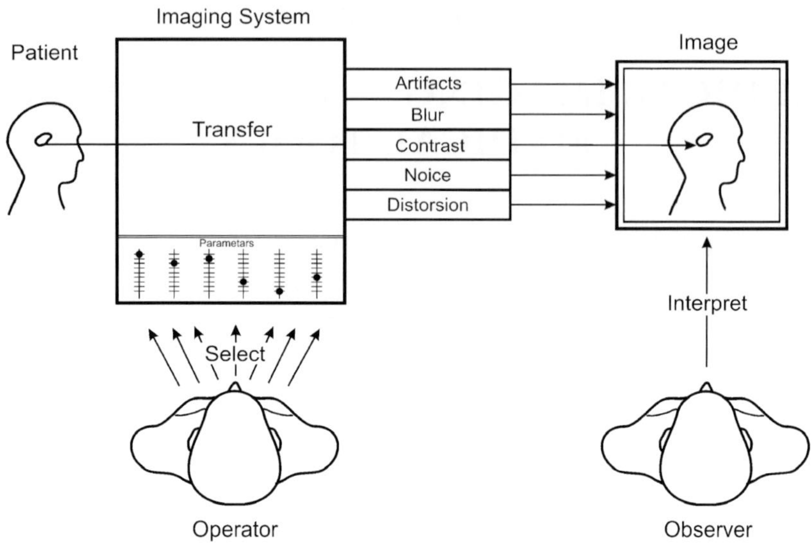

Figure 1. Components associated with medical imaging.[1]

on characteristics of the imaging equipment, skill of the operator, and compromises made with factors such as patient radiation exposure and imaging time.

Figure 1 is an overview of the medical imaging process. The five major components are the patient, imaging system, system operator, image itself and the observer. The aim is to make an object or condition within the patient's body visible to the observer. The visibility of specific anatomical features depends on characteristics of the imaging system and the manner in which it is operated. Most medical imaging systems have a considerable number of variables that must be selected by the operator. They can be changeable system components, such as intensifying screens in radiography, transducers in sonography, or coils in magnetic resonance imaging (MRI). However, most variables are adjustable physical quantities associated with the imaging process, such as *kilovoltage* in radiography, *gain* in sonography and *echo time (TE)* in MRI. The selected values will determine the quality of the image and visibility of specific body features.[1]

Advances in medical imaging have made it possible to get accurate information about human body condition quickly and easily. Examination may include traditional X-ray (radiography), fluoroscopy, computed tomography (CT), ultrasound, nuclear medicine, or magnetic resonance imaging (MRI). In addition to the Technologist, who administers your examination, and the Physician who interprets your images, Medical Physicist is also one of the caregivers. Medical Physicist is a member of the medical imaging team, ensuring patient images are of the highest quality. Medical Physicist is also the resource for the Physician and Technologist, helping them to understand better the technical aspects of imaging methods so they can use them most effectively.

2. Medical Imaging

2.1. *X-ray Images*

Since Roentgen's discovered that X-rays can identify bony structures, X-rays have been developed for use in medical imaging. Radiology is a specialised field of medicine. Radiographers employ radiography and other techniques for diagnostic imaging. Indeed, this is probably the most common use of X-ray technology.

All X-ray imaging is based on the absorption of X-rays as they pass through different parts of a patient's body. Different amount of X-rays will pass through and exit the body depending on the amount absorbed in a particular tissue such as muscle or lung. The amount of X-rays absorbed contributes to the radiation dose to the patient. During conventional X-ray imaging, the exiting X-rays interact with a detection device (X-ray film or other image receptor) and provide a 2-dimensional projection image of the tissues within the patient's body, see Figure 2.

X-rays are useful in particular in detection of the skeletal system pathology, but they are also useful for detecting some disease processes in a soft tissue. Some notable examples include highly common chest X-rays, which can be used to identify lung diseases such as pneumonia, lung cancer or pulmonary oedema, and the abdominal

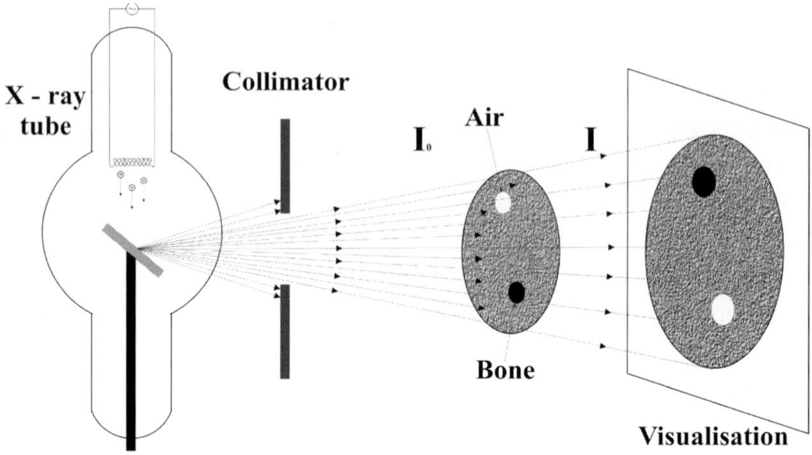

Figure 2. Scheme of application of X-rays in medical imaging: X-ray rube, Collimator, Body, X-ray image.

X-rays, which can detect ileum (blockage of the intestine), free air (from visceral perforations) and free fluid (in ascites). In some cases, the use of X-rays is debatable, such as gallstones (which are rarely radiopaque) or kidney stones (which are often visible, but not always). Also, traditional plain X-rays pose very little use in imaging of soft tissues such as brain or muscle. Imaging alternatives for soft tissues are computed axial tomography (CAT or CT scanning), magnetic resonance imaging (MRI) or ultrasound.

2.2. *CT Scan — Computed Tomography*

Computed Tomography (CT), originally known as computed axial tomography (CAT or CT scan) and body section rentenography, is a medical imaging method employing tomography where digital geometry processing is used to generate a three-dimensional image of the internals of an object from a large series of two-dimensional X-ray images taken around a single axis of rotation.

Since the first CT scanner was developed in 1972 by Sir Godfrey Hounsfield, the modality has become established as an essential radiological technique applicable in a wide range of clinical situations.

CT uses X-rays to generate cross-sectional, two-dimensional images of the body. Images are acquired by rapid rotation of the X-ray tube 360° around the patient. The transmitted radiation is then measured by a ring of sensitive radiation detectors located on the gantry around the patient, see Figure 3. The final image is generated from these measurements utilising the basic principle that the internal structure of the body can be reconstructed from multiple X-ray projections.[2]

Every acquired CT slice is subdivided into a matrix of up to 1024×1024 volume elements (voxels). Each voxel has been traversed during the scan by numerous X-ray photons and the intensity of the transmitted radiation is measured by detectors. From these intensity readings, the density or attenuation value of the tissue at each point in the slice can be calculated. The viewed image is then reconstructed as a corresponding matrix of picture elements (pixels).

The latest generation of CT scanners are multi-slice or multi-detector machines utilising the principles of helical scanner but incorporate multiple rows of detector rings. They can therefore acquire multiple slices per tube rotation, see Figure 3.[2]

2.3. *PET — Positron Emission Tomography*

Positron Emission Tomography (PET) is a radiotracer imaging technique, in which tracer compounds labelled with positron-emitting radionuclides are injected into the subject of the study. These tracer compounds track biochemical and physiological processes *in vivo*. One of the prime reasons for significance of PET in medical research and practice is the existence of positron-emitting isotopes of elements such as carbon, nitrogen, oxygen and fluorine, which may be processed to create a range of tracers, compounds which are similar to naturally occurring substances in the body.[3]

As the radioisotope undergoes positron emission decay (also known as positive beta decay), it emits a positron, the antimatter counterpart of an electron. After travelling up to a few millimetres the positron encounters and annihilates with an electron, producing a pair of annihilation (gamma) photons moving in almost opposite directions.

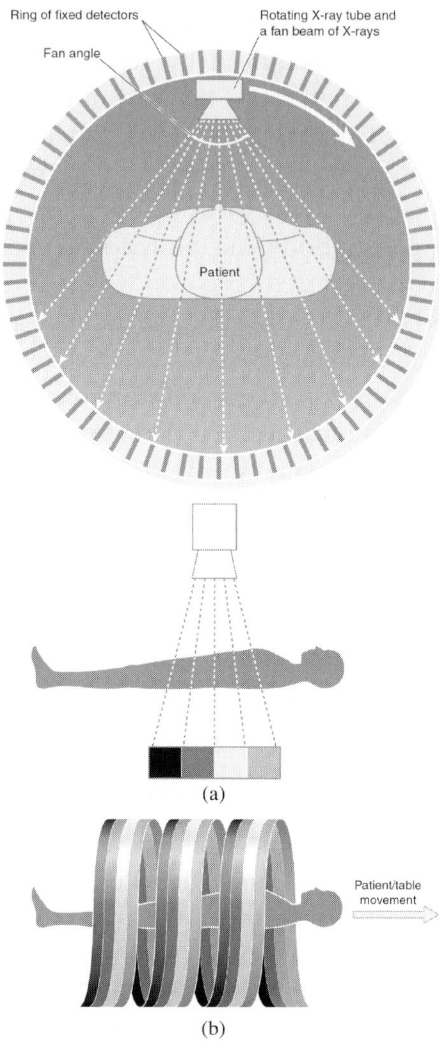

Figure 3. Ring of detectors (forth generation). Patient in cross-section (a) Multi-detector system (four rings shown here). (b) Multi-slice helical CT.[2]

Since two 511 keV gamma photons are emitted at almost 180 degrees to each other; it is possible to localise their source along a straight line of coincidence (also called formally the "line of response" or LOR), see Figure 4.

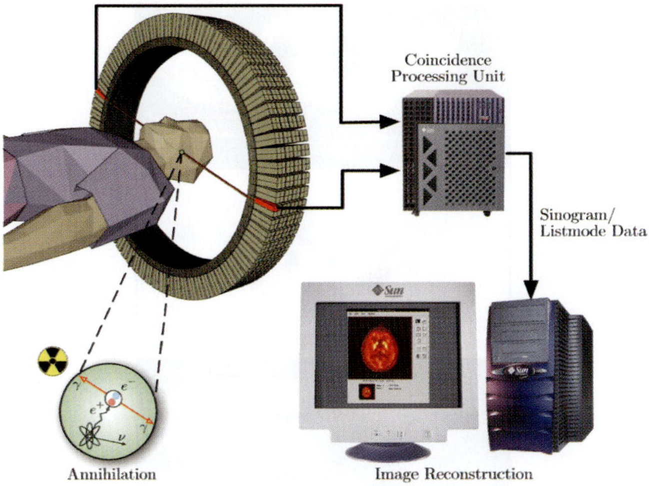

Figure 4. Basic steps in the image acquisition in the PET.

The raw data collected by a PET scanner are a list of "coincidence events" representing near-simultaneous detection of annihilation photons by a pair of detectors.

Coincidence events can be grouped into projections images, called sinograms. The sinogram images are analogous to the projections captured by computed tomography (CT) scanners, and can be reconstructed in a similar way. However, the statistics of the data is much worse than those obtained through transmission tomography.

The largest area of clinical use of PET is in oncology, cardiology and neurology. The most widely used tracer in oncology is ^{18}F–fluoro–deoxy–glucose (FDG) which follows a similar metabolic pathway to glucose *in vivo*, except that it is not metabolised to CO_2 and water, but remains trapped within a tissue. This makes it well suited to use as a glucose uptake tracer. This is of interest in oncology because proliferating cancer cells have a higher than average rate of glucose metabolism. ^{11}C–methionine is also used in oncology, where it acts as a marker for protein synthesis.[3]

Recently, scanners incorporating PET and computed tomography (CT) have become available, allowing sequential CT and PET imaging with the same device. The CT images can be efficiently co-registered

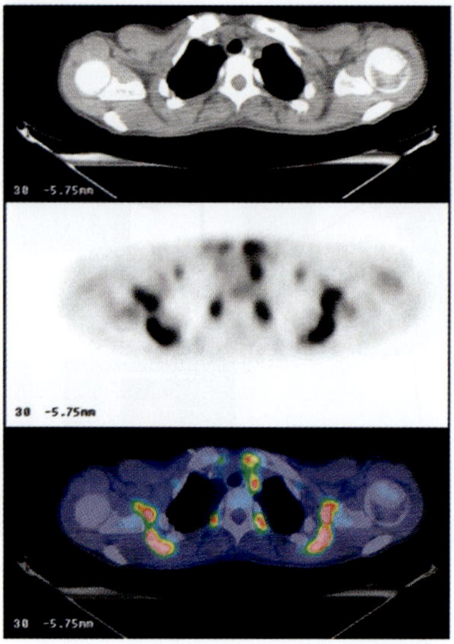

Figure 5. Shoulder girdles, superior mediastinum and paravertebral regions are shown in axial computed tomography (upper image), fluorine-18-fluorodeoxyglucose positron emission tomography (middle image) and fused PET/CT (bottom image) slices, demonstrating high metabolic activity in the low density fatty tissue (down).[4]

with the PET data so that the "fusion images" provide a structural template for simultaneous correlation with the metabolic image, see Figure 5. That this innovation promises to become the most comprehensive diagnostic tool in oncological imaging is confirmed by studies which have shown that these combined PET/CT scanners improve the accuracy of cancer evaluation compared with either device used alone.[4,5]

2.4. *MRI — Magnetic Resonance Imaging*

Magnetic resonance imaging (MRI) is a medical imaging process that uses a magnetic field and radio frequency (RF) signals to produce images of anatomical structures, of the presence of disease and of various biological functions within the human body.[6] The fundamental concept behind the MRI is the spinning nuclear charge, which produces

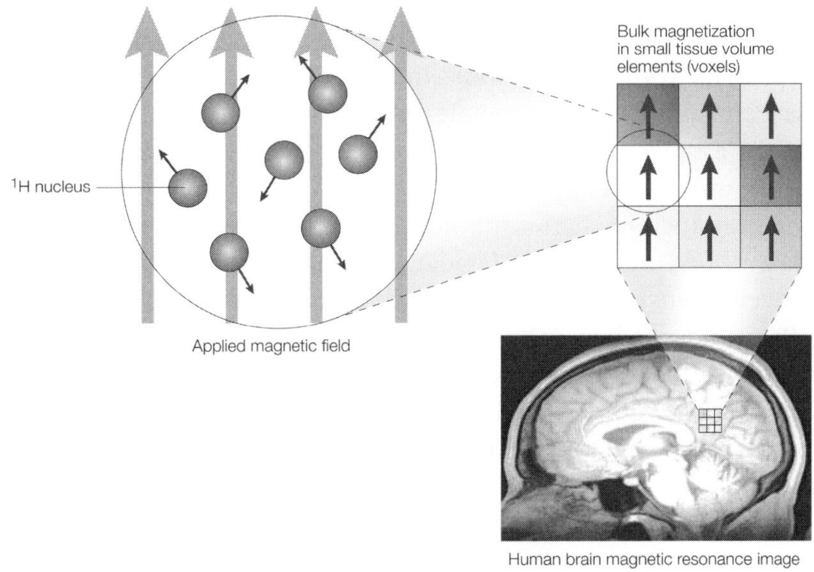

Figure 6. Principle of MRI with a typical image of the brain.

a tiny magnetic field. The simplest and most useful nucleus to consider is that of the hydrogen atom, naturally abundant in the human body and comprising a single proton. In the normal environment, the random orientation of millions of these spins results in no net magnetic field. However, the situation changes when these spins are placed in a strong external magnetic field (the scanners used in medicine have a typical magnetic field strength from 0.3 to 3 T), see Figure 6. The spins now align themselves in given directions, but rather than simply turning, their angular momentum causes the spins to precess at a certain frequency.[7]

The MR image is a display of RF signals that are emitted by the tissue during the image acquisition process. The source of the signals is a condition of magnetisation that is produced in the tissue when the patient is placed in the strong magnetic field. MRI should not be confused with the NMR (Nuclear Magnetic Resonance) spectroscopy technique used in chemistry, although both are based on the same principles of nuclear magnetism.

2.5. *Ultrasound Medical Imaging*

The primary form of ultrasonic imaging to date has been that of a pulse–echo mode. The principle is very similar to that of sonar and radar. In essence, following an ultrasonic pulse transmission, echoes from the medium being interrogated are detected and used to form an image. Much of the terminology used in ultrasound was imported from the field of sonar and radar. Although pulse–echo ultrasound had been used to diagnose a variety of medical problems since the 1950s, it did not become a widely accepted diagnostic tool until the early 1970s when grey-scale ultrasound with nonlinear echo amplitude to grey level mapping was introduced. Continuous wave (CW) and pulsed wave Doppler (PW) ultrasound devices for measuring blood flow also became available during that time. Duplex ultrasound scanners that combined both functions, thus allowing the imaging of anatomy and the measurement of blood flow with one single instrument, soon followed. The rapid progress of computer science has enabled applications known as 3D, and 4D ultrasound. With these ultrasound applications we can acquire 3D images, even watch real-time moves of the object. Today, ultrasound is the second most utilised diagnostic imaging modality in medicine, second only to conventional X-ray, and is a critically important diagnostic tool of any medical facility, see Figure 7.[8]

The devices, which both create the sound and listen for the echoes, use transducers made of piezoelectric materials. These materials have a wonderful property of vibrating when stimulated by electricity and of generating electricity when they are vibrated by sound. Electric pulses created in transducers by the echoes are converted to digital representations which can then be converted into anything from a simple string of numbers to a picture on a video monitor. Once the reflections of the sound waves have been captured by a machine, the machine then transforms the reflections into an image which can be read.

Ultrasound is an art which does not involve the use of radiation or any type of dyes that need to be injected or swallowed. It is safe, fast and very effective for certain types of medical imaging.

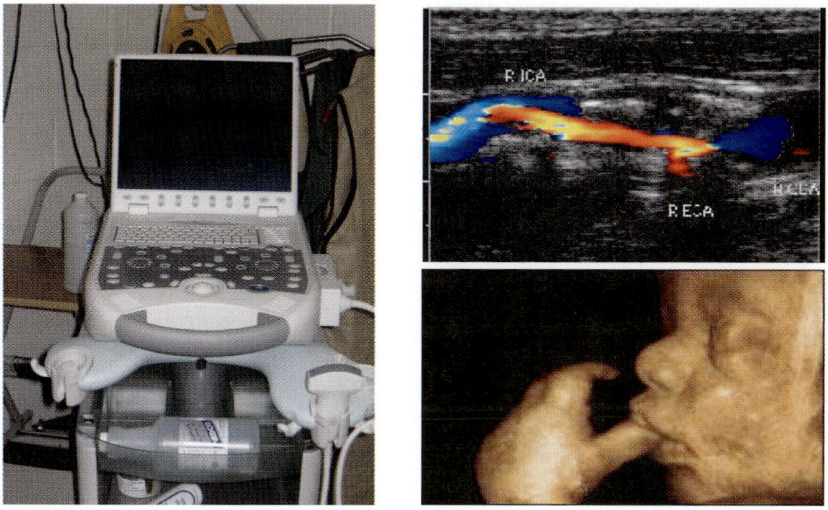

Figure 7. Ultrasound device (left), ultrasound image of the blood flow — Doppler technique (up-right), 3D ultrasound image of an unborn baby (down-right).

Although ultrasound has been in existence for more than 40 years and is considered a mature technology by many, the field is by no means in a stagnant state. Technical advances are still being made. The introduction of contrast agents, harmonic imaging, flow and tissue displacement imaging, multidimensional imaging and high-frequency imaging are just a few examples.

References

1. P. Sprawls, *Physical Principles of Medical Imaging* (Medical Physics Publishing, 1995).
2. S.A. Jackson and R. Thomas, *Cross-sectional Imaging Made Easy* (Elsevier Health Sciences, 2004).
3. V. Spasić-Jokić, Positron Emission Tomography (PET) in medical imaging, TESLA Accelerator Installation, Laboratory of Physics, VINCA Institute of Nuclear Sciences, Belgrade (2007).
4. W.F.E. Lau, D.S. Binns, R.E. Ware, S. Ramdave, F. Cachin, A.G. Pitman and R.J. Hicks, *The Med. J. Australia* **182**, 172 (2005).
5. H. Schoder, S.M. Larson and W.D. Yeung, *J. Nucl. Med.* **45**, 72S (2004).

6. P. Sprawls, *Magnetic Resonance Imaging* (Medical Physics Publishing, 2000).
7. G. Liney, *MRI in Clinical Practice* (Springer-Verlag London Limited, 2006).
8. K.K. Shung, *Diagnostic Ultrasound* (Taylor & Francis Group, 2006).

CHAPTER 2

POSITRON EMISSION TOMOGRAPHY (PET) IN MEDICAL IMAGING

VESNA SPASIĆ-JOKIĆ

Laboratory of Physics, "VINČA" Institute of Nuclear Sciences, Belgrade, Serbia

Positron Emission Tomography (PET) is a radiotracer imaging technique, in which tracer compounds labelled with positron-emitting radionuclides are injected into the subject of the study. One of the prime reasons for the importance of PET in medical research and practice is the existence of positron-emitting isotopes of elements such as carbon, nitrogen, oxygen and fluorine, which may be processed to create a range of tracer compounds which are similar to naturally occurring substances in the body. The largest area of clinical use of PET is in oncology, cardiology and neurology. The most widely used tracer in oncology is ^{18}F–fluoro–deoxy–glucose (FDG) which follows a similar metabolic pathway to glucose *in vivo*, except that it is not metabolised to CO_2 and water, but remains trapped within a tissue. This makes it well-suited to use as a glucose uptake tracer. This is of interest in oncology because proliferating cancer cells have a higher than average rate of glucose metabolism. ^{11}C-methionine is also used in oncology, where it acts as a marker for protein synthesis. In our country, one of the experimental channels of TESLA Accelerator Installation will be used for production of long-living positron emitters (e.g. ^{124}I, ^{86}Y, ^{76}Br, ^{64}Cu) and alpha emitters (e.g. ^{211}At and ^{149}Tb).

1. Introduction

1.1. *What is PET?*

Positron Emission Tomography (PET) is a radiotracer imaging technique, in which tracer compounds labelled with positron-emitting radionuclides are injected into the subject of the study. These tracer compounds track biochemical and physiological processes *in vivo*. One of the prime reasons for the importance of PET in medical research and practice is the existence of positron-emitting isotopes of elements such as carbon, nitrogen, oxygen and fluorine which may be processed to create a range of tracer compounds which are similar to naturally occurring substances in the body.

While PET was originally used primarily as a research tool, in recent years it has come to have an increasingly important clinical role. The largest area of clinical use of PET is in oncology. The most widely used tracer in oncology is ^{18}F–fluoro–deoxy–glucose (^{18}F–FDG). ^{18}F–FDG is relatively easy to synthesise with a high radiochemical yield.[1] It also follows a similar metabolic pathway to glucose *in vivo*, except that it is not metabolised to CO_2 and water, but remains trapped within tissue. This makes it well-suited to use as a glucose uptake tracer. This is of interest in oncology because proliferating cancer cells have a higher than average rate of glucose metabolism. ^{11}C-methionine is also used in oncology, where it acts as a marker for protein synthesis.

1.2. *Application of PET*

During the last decade, PET imaging has had a progressive impact on clinical medicine, dating approximately from the April 1991 issue of the *Journal of Nuclear Medicine* which announced that "Clinical PET had arrived". Clinical PET services have grown rapidly and are now widely used throughout clinical medicine, in particular in oncology. This expansion continues and PET is now the fastest growing area in diagnostic imaging.

In Oncology PET is very confident in Confirmation and Staging of suspected malignancy. As cancer has higher glycolytic rates than most normal tissues, FDG–PET can be used to differentiate benign from malignant tissue and to a limited extent define the grade of malignancy and hence the prognosis. Brain tumours were the first tumours to be studied by FDG–PET and it has been well established that it is possible to confirm malignancy and establish grade and prognosis. It may be most valuable for primary brain tumours when biopsy is difficult or carries too great a risk. This is often particularly important in relation to cerebral masses in HIV/AIDS patients.[2]

Diagnosis of recurrent disease: FDG–PET can contribute to management of variety of cancers by confirming treatable recurrent disease. The reasons for this are the high sensitivity of FDG–PET and the fact that after surgery or radiation therapy the normal anatomy may be distorted so that standard anatomical imaging methods may be extremely difficult to interpret added to which it may not be possible to distinguish scars from tumour.

It may be difficult to identify the site of a primary malignant tumour even when there is a very high probability of disease being present. FDG–PET has been found to be useful for identifying the site of squamous cell, head and neck carcinoma when a patient displays the presence of nodes and the primary site is unknown (CUP syndrome), for identifying the site of tumour in patients who are suspected of having para neo-plastic tumours and for patients with raised markers, including colorectal cancer with carcino-embryonic antigen (CEA) marker α-Feto proteins (α-FP) for patients with teratoma. Serum thyroglobulin and calcitonin patients with differentiated follicular and medullar thyroid carcinoma are further examples for which FDG–PET scans have been demonstrated to be of clinical benefit for tumour localisation.[3–5]

FDG–PET may also have a role in defining the limits of a tumour for radiotherapy or surgical excision when defining the line of demarcation between tumour and inflammatory response is not possible using anatomical imaging. Head and neck tumours are probably the best examples. When biopsy is required at a particular part of a tumour with, for example, the site of the highest stage disease, FDG–PET may be used to direct biopsy; mesothelioma and sarcoma are two good examples as there may be a highly considerable non homogeneity in disease activity throughout the tissue.

FDG–PET is increasingly being used to measure and predict the response to therapy (radiation or chemotherapy) and this may be the most important role for PET in future, possibly with other tracers, and this is discussed in more depth later. FDG–PET can distinguish residual active from inactive masses in lymphoma and teratoma and a recent study demonstrated a high positive predictive accuracy for lymphoma relapse (100% FDG) *versus* (41%) for CT. Increasingly, FDG–PET is being used to measure response to therapy in breast cancer, ovarian cancer and for neo-adjuvant chemoradiation when it is used for downstaging prior to surgical excision (e.g. rectal, oesophageal and lung cancers) but prospective studies are needed to determine the best way to use this information.

Proton Therapy Treatment Verification via PET Imaging is used in many cases such as early estimation of tumour therapy effects which can show suppression after two cycles of chemo or radiotherapy even when there is no visible reduction of tumour mass; early determination of resistant response on therapy aimed at avoiding toxic effects of inefficient therapy; FDG can divide benignant from malignant tumours in order to avoid expensive surgery intervention; estimation of myocardial metabolism by FDG enable high risks identification without using coronary invasive techniques and decrease the health insurance costs. Therapy verification can be achieved by comparing the PET images with the predicted target dose distribution used in planning obtained by MC. We developed our own SRNA package (original author Dr R. Ilic).[2–4]

Although clinical PET has become dominated by oncological applications, which are also increasing, there is also a small but important role of it in neuropsychiatry. The first is accompanying the routine work up of patients having elective surgery for the control of partial epilepsy. Often there are no problems in identifying the culprit focus, usually in the temporal lobe with the MRI showing mesial temporal sclerosis that can be lateralised with EEG recordings. When, however, there is no clear-cut or the evidence is conflicting, an interictal FDG–PET brain scan may be crucial in demonstrating lateralisation of hypo metabolic changes. Alternatively, it may be helpful if the focus is suspected to be either multi-focal or outside the temporal lobes, often in conjunction with ictal single photon emission computed tomography (SPECT) studies which are not possible with PET. Changes in FDG–PET brain scan have been shown to be the earliest objective finding for the diagnosis of dementia and differential diagnosis of the type of dementia. As treatments for Alzheimer's disease become available, this role is likely to increase significantly.

Application of PET in cardiology is possible by using NH_3 for myocardial perfusion with adenosine. For coronary vasodilatation it is the most accurate method of diagnosing the functional effects of coronary artery disease (CAD). However, because of the cost and its limited availability it is not cost effective to use PET for routine diagnosis of CAD, although it remains valuable in particularly difficult cases. It is certainly possible to diagnose myocardial ischemia with a high sensitivity and specificity, but it is not generally appropriate.

What are the future directions of clinical PET? It does appear as though further work with software image registration and integrated PET/CT systems is going to be an important part of the extension of this imaging modality. Huge opportunities will be presented by new ligands and the new concepts of molecular imaging in many diseases and for monitoring gene therapies. In cancer, rather than current staging methods based on disease spread it is likely that biological characterisation of a tumour will become much more important in selecting patients for particular therapies, and PET will have the key role.[3–5]

2. Positron Emission and Annihilation

Proton-rich isotopes may decay via positron emission, in which a proton in the nucleus decays to a neutron, a positron and a neutrino. The daughter isotope has an atomic number that is one less than the parent. Examples of isotopes which undergo decay via positron emission are shown in Table 1.

As positrons travel through a human tissue they give up their kinetic energy principally by Coulomb interactions with electrons. As the remaining mass of the positron is the same as that of the electron,

Table 1. Properties of commonly used positron emitting radio-isotopes[6,7]

Isotope	Half-life (min)	Maximum positron energy (MeV)	Positron range in water (FWHM in mm)	Production method
^{11}C	20.3	0.96	1.1	cyclotron
^{13}N	9.97	1.19	1.4	cyclotron
^{15}O	2.03	1.70	1.5	cyclotron
^{18}F	109.8	0.64	1.0	cyclotron
^{68}Ga	67.8	1.89	1.7	generator
^{82}Rb	1.26	3.15	1.7	generator

the positrons may undergo large deviations in direction with each Coulomb interaction, and they follow a tortuous path through the tissue as they give up their kinetic energy.

When positrons reach thermal energies, they start to interact with electrons either by annihilation, which produces two 511 keV photons which are anti-parallel in the positron's frame, or by the formation of a hydrogen-like orbiting couple called positronium. In its ground-state, positronium has two forms — ortho-positronium, where the spins of the electron and positron are parallel, and para-positronium, where the spins are anti-parallel. Para-positronium again decays by self-annihilation, generating two anti-parallel 511 keV photons. Ortho-positronium self-annihilates by the emission of three photons. Both forms are susceptible to the "pick-off" process, where the positron annihilates with another electron. Free annihilation and the pick-off process are responsible for over 80% of the decay events. Variations in the momentum of interacting particles involved in free annihilation and pick-off result in an angular uncertainty in the direction of the 511 keV photons of around 40 μGy in the observer's frame. In a PET camera with 1 m of diameter and active transaxial FOV of 0.6 m this results in a positional inaccuracy of 2–3 mm.

The final positron range and the non-collinearity of annihilation photons give rise to an inherent positional inaccuracy not present in conventional single photon emission techniques. However, other characteristics of PET which are discussed below more than offset this disadvantage.

3. Coincidence Detection and Electronic Collimation

In a PET camera, each detector generates a timed pulse when it registers an incident photon. These pulses are then combined in coincidence circuitry, and if the pulses fall within a short time-window, they are deemed to be coincident.

A coincidence event is assigned to a line of response (LOR) joining the two relevant detectors. In this way, positional information is gained from the detected radiation without the need for a physical collimator. This is known as *electronic collimation*. Electronic collimation has two major advantages over physical collimation. These are improved sensitivity and improved uniformity of the point source response function.

When a physical collimator is used, directional information is gained by preventing photons which are not normal or nearly normal to the collimator face from falling on the detector. In electronic collimation, these photons may be detected and used as signal. This results in a significant gain in sensitivity (typically a factor of 10 for 2D mode PET compared with SPECT). This increase in sensitivity means that typical realisable image resolution in PET is around 5–10 mm, whereas in SPECT it is around 15–20 mm. Coincidence events in PET fall into 4 categories: true, scattered, random and multiple.

True coincidences occur when both photons from an annihilation event are detected by detectors in coincidence, neither photon undergoes any form of interaction prior to detection and no other event is detected within the coincidence time-window.

A scattered coincidence is one in which at least one of the detected photons has undergone at least one Compton scattering event prior to detection. Since the direction of the photon is changed during the Compton scattering process, it is highly likely that the resulting coincidence event will be assigned to the wrong LOR. Scattered coincidences add a background to true coincidence distribution which changes slowly with position, decreasing contrast and causing the isotope concentrations to be overestimated. They also add statistical noise to the signal. The number of scattered events detected depends on the volume and attenuation characteristics of the object being imaged and on geometry of the camera.

Random coincidences occur when two photons not arising from the same annihilation event are incident on the detectors within the coincidence time window of the system. The number of random coincidences in a given LOR is closely linked to the rate of single events measured by the detectors joined by that LOR and the rate of random coincidences increase roughly with the square of the activity in the FOV. As with scattered events, the number of random coincidences detected also depends on the volume and attenuation characteristics of the object being imaged and on geometry of the camera. The distribution of random coincidences is fairly uniform across the FOV and it will cause isotope concentrations to be overestimated if not corrected for. Random coincidences also add statistical noise to the data.

Multiple coincidences occur when more than two photons are detected in different detectors within the coincidence resolving time. In this situation, it is not possible to determine the LOR to which the event should be assigned, and the event is rejected. Multiple coincidences can also cause the event of mispositioning.

4. Safety Aspects

PET scanning is non-invasive technique, but it does involve exposure to ionizing radiation. Total dose per procedure is estimated as around 7 mSv.[8–12] Bladder wall receives 0.12 mGy/MBq; Heart could receive 0.05 mGy/MBq. Brain receives around 0.03 mGy/MBq and other organs receive (0.01 to 0.03) mGy/MBq. Estimated effective dose is 0.024 mSv/MBq (around 10 mSv per procedure). PET emitters in Serbia will be produced at TESLA Accelerator Installation in VINČA Instititute of Nuclear Sciences, Laboratory of Physics.

Programmes at TESLA include: modification of materials by ion beams; radiation researches; physics of thin crystals and nanotubes; production of radioisotopes and radiopharmaceuticals; proton therapy;

neutron research and physics of hadrons and electroweak interactions at medium and high energies. Plans for Production of Radionuclides and Radiopharmaceuticals at TAI consist of three groups of products:

1. Radiopharmaceuticals for routine nuclear medical diagnostic procedures based on ^{18}F; ^{123}I; ^{201}Tl; ^{67}Ga; ^{111}In. Production plan predicts 1 GBq FDG per year which is enough for 2 300 diagnostic procedures in Clinical Centre of Serbia in Belgrade, Clinical Centre in Sremska Kamenica and in a few private hospitals,
2. Radiopharmaceuticals for pre-clinical and clinical tests based on alternative positron emitting radionuclides: ^{124}I; ^{86}Y; ^{64}Cu,
3. Radionuclides for alpha-endoradiotherapy: ^{211}At; ^{149}Tb.

References

1. K. Hamacher, H.H. Coenen and G. Stocklin, *J. Nucl. Med.* **27**, 235 (1986).
2. R.D. Ilić, V. Spasić-Jokić, P. Beličev and M. Dragović, *Phys. Med. Biol.* **50**, 1011 (2005).
3. V. Spasić-Jokić, Treatment planning in proton therapy, Medical physics in South-Eastern Europe, *EFOMP International School: Treatment Planning Systems*, Fruska Gora, Serbia (2004).
4. V. Spasić-Jokić, Protocol for proton dosimetry, Monography: Cyclotrons in medicine, 285 (1993).
5. V. Spasić-Jokić, *Sci. Technol. Securit.* **2**, 31 (1996).
6. R.A. Raylman, B.E. Hammer and N.L. Christensen, *Trans. Nuc. Sci.* **43**, 2406 (1993).
7. D.L. Bailey, Quantification in 3D positron emission tomography, PhD Thesis, University of Surrey (1996).
8. M. Orlić, M. Jovanović, V. Spasić-Jokić, O. Čuknić. Ž. Ilić and S. Vranješ, Additional effective dose by patients undergoing NaI-131 capsules therapy, *2nd European IRPA Congress on Radiation Protection: From knowledge to action,* France, *Proceeding of full papers, CD Rom*, 1011 (2006).
9. B. Đurović, V. Selaković and V. Spasić-Jokić, *Arch. Oncolog.* **12**, 197 (2004).

10. V. Spasić-Jokić, M. Orlić and B. Djurović, Risk of potential radiation accidental situations as TESLA Accelerator Installation, *2nd European IRPA Congress on Radiation Protection: From knowledge to action,* France, *Proceeding of full papers, CD Rom* (2006).
11. B. Đurović, V. Spasić-Jokić and M. Mišović, The role of radiologic technologist in radiation protection and quality assurance programs, *3rd International Yugoslav Nuclear Society Conference, YUNSC-2000,* Serbia, *Proceedings,* 700 (2000).
12. M. Orlić, V. Spasić-Jokić, M. Jovanović and S. Vranješ, Calibration of nuclear medicine gamma counters, *5th International Yugoslav Nuclear Society Conference, YUNSC-2004, Beograd Proceedings*, Serbia, 405 (2004).

CHAPTER 3

DICOM/JPEG2000 CLIENT/SERVER ARCHITECTURE IMPLEMENTATION

DINU DRAGAN AND DRAGAN IVETIĆ

Faculty of Technical Sciences, University of Novi Sad, Novi Sad, Serbia

The paper presents client/server architecture based on DICOM standard (Digital Imaging and Communications in Medicine). Standard DICOM transfer syntax is extended to support JPEG2000 streaming. This grants DICOM clients ability to query and retrieve medical images in resolution and quality suitable to their display capabilities which is very useful when DICOM client runs on a device which has smaller display resolution, like palmtop computers or mobile phones. The benefits of this client/server architecture are analysed. The paper presents client/server prototype based on the extended DICOM transfer syntax from the end-user's perspective. The applications based on these prototypes are meant for use in Hospital for Lung Diseases "Dr Vasa Savić Zrenjanin".

1. Introduction

Picture archiving and communication system (PACS) is a workflow-integrated imaging system designed to streamline the operations throughout the entire patient care delivery process.[11] It facilitates digital medical image communication aspect of the electronic healthcare. Although it may have a broadened meaning, the term PACS usually refers to a computer based system designed to store, retrieve and display medical images.[10] The architecture of typical PACS is shown in Figure 1.

DICOM (Digital Imaging and Communication in Medicine) communication protocol is the interface used normally for communication

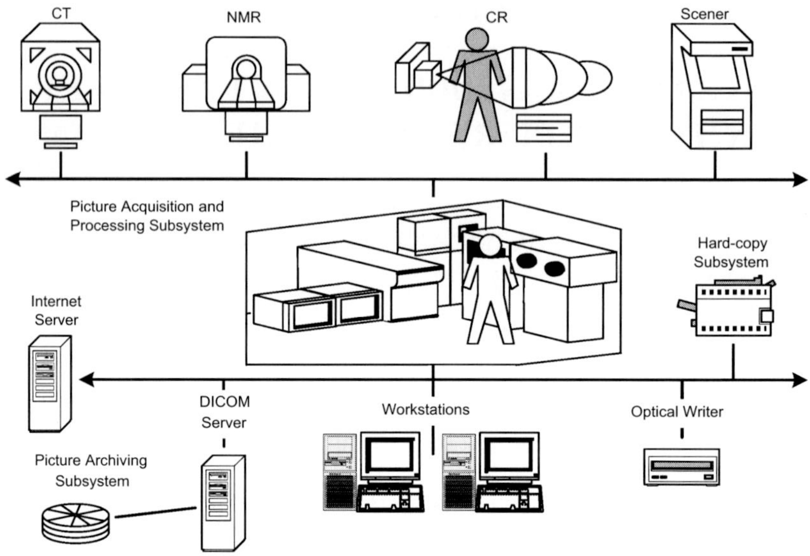

Figure 1. Illustration of the picture archiving and communication system.

between different parts of the PACS.[1] The DICOM standard specifies a non-proprietary data interchange protocol, digital image format and file structure for biomedical images and image-related information.[8,17] It is not the only standard for communication of medical related messages, but in the domain of medical imaging it has an "unparalleled international success".[7]

One of the major issues of the PACS development and management are requirements for large storage capacities and high-speed networks. As an example, PACS in a general medical centre requires about 40 Terabytes of storage capacity for seven years and network that supports at least 100 Mbits/s data flow.[11,12] These requirements exist because digital medical images tend to be very large. For example, a single analogue mammogram might be digitised at 4096 × 4096 pixels × 16 bpp which means that this file would be about 33 megabytes (MB).[14] The only way to address this problem is by using various techniques for image compression. Image compression can reduce the storage and bandwidth requirements for delivering images to clinicians.

The JPEG2000 is one of the latest compression standards and represents the state of the art in the field of still image compression. It achieves good compression rates with the minimum space distortions. It also provides many features that support scalable and interactive access to large images.[2,15] These include efficient lossless and lossy compression, resolution scalability, quality scalability, region of interest coding and spatial random access. Some research of JPEG2000 compression of medical images has been performed and reported favourable image quality evaluations.[4–6]

In this paper, we propose client/server architecture, named DICOM2000 system, based on the extended DICOM transfer syntax which supports JPEG2000 streaming. This client/server architecture is suitable for PACS. Extended DICOM syntax enables the use of JPEG2000 resolution scalability, quality scalability and spatial random access through DICOM communication channels. The proposed changes are transparent for DICOM based PACS, which enables interoperability with programmes that implement non-extended DICOM transfer syntax.

The organisation of the paper is as follows: Section 2 describes the architecture of the DICOM2000 system and gives a description of extensions made on standard DICOM syntax, Section 3 describes the benefits of the DICOM2000 system in the clinician's practices and Section 4 provides some concluding remarks.

2. The DICOM2000 System

There are several possible architectures for building PACS based on DICOM standard and JPEG2000 compression. Some of these architectures support DICOM standard but they facilitate only the basic features described by DICOM standard while JPEG2000 streaming is not one of them. Other architectures implement PACS based on extended DICOM transfer syntax. One of these architectures is described later in this section, see Figure 2. The description of all possible architectures of PACS based on DICOM standard and JPEG2000 compression will be addressed elsewhere.[9]

2.1. *DICOM2000 Architecture*

Figure 2 describes PACS architecture based on extended DICOM transfer syntax and JPEG2000 compression. This architecture enables the use of JPEG2000 resolution scalability, quality scalability and spatial random access through DICOM communication channels.

The PACS architecture described in Figure 2, is client/server architecture. It contains two servers. One of them is DICOM server used to communicate with DICOM clients. These clients can be PC computers, LAPTOP computers, palmtop computers and/or mobile phones. The server and clients communicate through DICOM protocol (messages).

Another server in the architecture described in Figure 2, is JPEG2000 server which communicates through JPIP protocol (JPEG2000 Interactive Protocol). When communicating to JPEG2000 server, DICOM server is acting like JPIP client. The data associated with medical images are stored on the DICOM server while medical images are stored in JPEG2000 format on JPEG2000

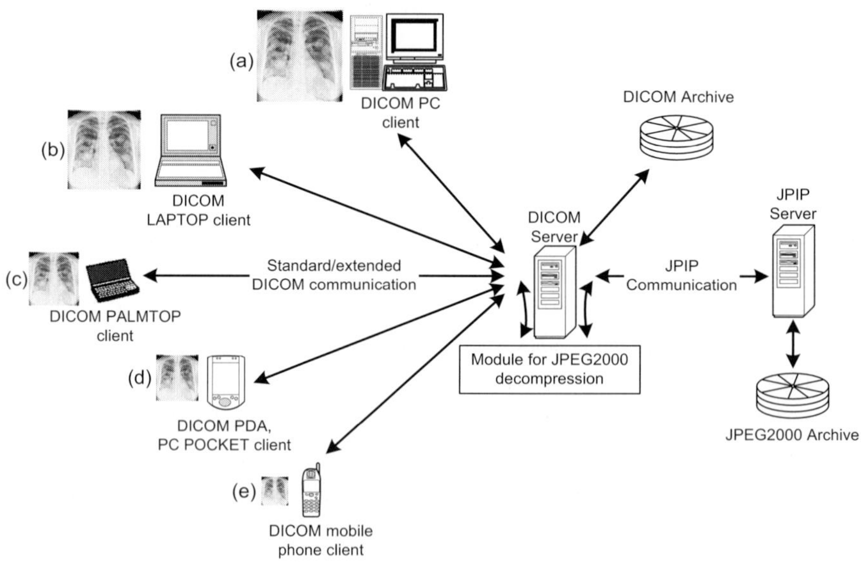

Figure 2. DICOM2000 Architecture.

server. JPIP protocol is the ninth part of the JPEG2000 standard. It describes protocol for JPEG2000 images interchange. The purpose of JPIP is to standardise the means of interacting with JPEG2000 based data in an efficient and effective manner.[16]

To be fully DICOM compliant, DICOM server from Figure 2 implements both standard and extended DICOM transfer syntax through the mechanism described in DICOM standard.[1,8] To fulfil this requirement DICOM server has to be able to decompress JPEG2000 images and to transcode them into DICOM original format. This means that any DICOM conformant application can communicate with this DICOM server.

The extended DICOM transfer syntax enables compliant DICOM clients to define their viewing resolution, image quality and image viewing region (part of the image they want to view), when querying DICOM server. DICOM server processes these demands and sends them further to JPEG2000 server. The JPEG2000 server transcodes the desired JPEG2000 stream according to client's demand and streams it back to DICOM server. At the DICOM server side, the received JPEG2000 stream is encapsulated inside DICOM message along with associated data and sent back to DICOM client. Encapsulated JPEG2000 stream is extracted, decompressed and presented on the client's display.

Medical images in Figure 2 are represented in resolution and quality suitable for corresponded display devices.

2.2. *DICOM2000 Extensions to Standard DICOM Transfer Syntax*

Beside its original format, DICOM standard defines encapsulated message format that enables encapsulation of other digital image formats. JPEG2000 compression format is one of the formats supported by DICOM standard. This means that pixels inside the DICOM message are represented as JPEG2000 data stream. Unfortunately, this does not mean that client can request images of precise resolution, quality and/or precise part of the images. It only means that entire JPEG2000 stream has to be encapsulated into DICOM message regardless client's display resolution and memory capacities and that its transport and storage are not affected.

DICOM standard defines the mechanism for extending its transfer syntax, which enables the use of custom DICOM messages. In DICOM2000 system, DICOM message sent by DICOM client is extended with data fields containing information about requested image resolution, quality and requested part of the image. These fields are formatted according to JPIP protocol.[2] For example, if a client requires certain part of an image in certain resolution, it has to define next three data fields: resolution value defined as pair of values (width and height), image part dimension defined as value pair (width and height) and coordinates of the image part in the overall image defined as value pair (x and y coordinates).

The extended DICOM message sent by DICOM server contains response from JPEG2000 server which describes the image served to a client. These fields are also formatted according to JPIP protocol.[2] Server response fields are needed because JPEG2000 server is not obligated to reply entirely to a client's request, which means that the attributes of the image served are as close to the requested as possible. For the previous example, server response would contain the information of the actual resolution, image part and position of the served image.

3. Advantages of the DICOM2000 System

From the user's point of view, DICOM2000 system provides several benefits. PC based DICOM client application has the ability to preview a collection of medical images. This is useful in situations when a user may initially only need image thumbnails to decide which image he/she wants to view. In standard DICOM system clients do not have the possibility of image overview. The only way to find out if image is the one they need is to obtain it completely from the server. In low-band networks this is time-consuming. Without these extensions the only other way to implement image previewing is by using another communication system parallel to DICOM based PACS. Contrary to these systems, DICOM2000 system provides this feature through DICOM communication channel.

In client applications with small size display device, a low-resolution image may be delivered from the server. The initial resolution may be defined by the user or it can be completely transparent to the user and automatically extracted. In the second case, the image resolution is taken in accordance with the display device resolution. This is very useful when clients are on palmtop computers or mobile phones. It is also important to note that these devices usually communicate through low-band networks.

The user may define region of medical image that is of interest to her/him and only the data needed to display that region is sent to the client. For client applications with limited size display device this means that user may view the region of interest even in higher resolution, enabling the clinician to make some short-term diagnosis. In PC based client application, user may mark that region graphically on a previewed image. The information is collected and then sent to the server. This is important for low-band networks because it enables the user to view the region of interest very quickly without the need to wait for the rest of the image.

The DICOM2000 system has two drawbacks. First, which is easily overlooked, is that medical image always has to be accompanied by associated DICOM data. In interactive image exchange these data are sent with every new version of the image although it remains the same. This problem can be overlooked because these data are small in size (smaller then 7 kilobytes). The other one is that the DICOM2000 does not implement the JPEG2000 streaming entirely. Because JPEG2000 stream is encapsulated in DICOM message, the image cannot be presented progressively.

In our research, we implemented DICOM2000 system. The implementation consists of one server application which implements DICOM server and two client applications implementing PC and mobile phone part described in Figure 2. The problem of displaying images encapsulated in DICOM message on mobile devices is addressed in.[13] Kakadu JPEG2000 server is used as JPEG2000 server, see Figure 2.[3] The PC based client application achieves all the benefits previously described which were proved through its testing.

The DICOM2000 system developed is a part of the research made for the project titled *E-roentgenology of Special Hospital for Lung Disease "Dr. Vasa Savić Zrenjanin"* No 6233B, financed by the Government of the Republic of Serbia. Further testing of the system is going to take place there.

4. Conclusion

In this paper we described client/server architecture, named DICOM2000 system, based on the extended DICOM transfer syntax. DICOM standard is a communication protocol which enables communication of various parts/departments within a healthcare facility PACS. We extended DICOM transfer syntax to support JPEG2000 streaming. By doing this we enabled the use of JPEG2000 resolution scalability, quality scalability and spatial random access through DICOM communication channels.

The DICOM2000 system provides users with PC based client application to preview medical images and to view only the region of an image that is of interest to them. This feature of the DICOM2000 system enables fast image viewing even in low-band networks. It also enables viewing of DICOM data and low-resolution medical images on a limited size display devices. The second feature of the DICOM2000 system allows clinicians to view medical images at site, outside their hospitals, but yet inside DICOM network.

The initial testing of DICOM2000 system produced good and expected results. Since the DICOM2000 system is developed as a part of the research made for the project titled *E-roentgenology of Special Hospital for Lung Disease "Dr. Vasa Savić Zrenjanin"* No 6233B, financed by the Government of the Republic of Serbia, further testing of the system is planned to take place there.

References

1. DICOM, http://medical.nema.org/
2. JPEG2000, http://www.jpeg.org/jpeg2000
3. Kakadu Software, http://www.kakadusoftware.com/

4. A. Przelaskowski, Compression of mammograms for medical practice, *Proceedings of the 2004 ACM Symposium on Applied Computing*, 249 (2004).
5. B.J. Erickson, Irreversible compression of medical images, A White Paper from Society for Computer Applications in Radiology (SCAR), (2000).
6. D.H. Foos, E. Muka, R.M. Slone, B.J. Erickson, M.J. Flynn, D.A. Clunie, L. Hildebrand, K. Kohm and S. Young, JPEG2000 compression of medical imagery, *Proceedings of the SPIE, PACS Design and Evaluation*, 3980 (2000).
7. D.A. Clunie, *DICOM Structured Reporting*, (PixelMed Publishing, 2000).
8. D. Dragan, D. Ivetić, DICOM overview, *7th International Symposium "Young People and Multidisciplinary Research"*, Resita, Romania (2005).
9. D. Dragan, Encapsulation of JPEG2000 Compression by DICOM standard message, Author reprint, Library of Faculty of Engineering, Novi Sad, Serbia (2000).
10. E.E. Siegel, PACS, Current technology requirements and future trends, *Proceedings of the 4th International Conference on Image Management and Communications*, 2 (1995).
11. H.K. Huang, *Comp. Med. Imag. Graph.* **27**, 241 (2003).
12. H.K. Huang, A. Zhang, B. Liu, Z. Zhou, J. Documet, N. King, and L.W.C. Chan, Data grid for large-scale medical image archive and analysis, *Proceedings of the 13th annual ACM International Conference on Multimedia MULTIMEDIA '05* (2005).
13. J. Mirković, D. Ivetić and D. Dragan, Presentation of medical images extracted from DICOM objects on mobile devices, *In preparation for the 9th International Symposium of Interdisciplinary Regional Research "ISIRR 2007" Hungary–Serbia–Romania*, Novi Sad, Serbia (2007).
14. J. Strom and P. Cosman, **59**, 155 (1997).
15. D. Taubman, and M. Marcellin, *JPEG2000: Image Compression Fundamentals, Standards and Practice*, **795** (Kluwer Academic Publishers, 2000).
16. D. Taubman and R. Prandolini, Architecture, Philosophy and Performance of JPIP: Internet Protocol Standard for JPEG2000, *International Symposium on Visual Communications and Image Processing (VCIP2003)*, SPIE 5150, 791 (2003).
17. W.D. Bidgood, *J. Amer. Medic. Informatics Assoc.* **4**, 199 (1997).

CHAPTER 4

RADIOACTIVE ISOTOPES IN DIAGNOSIS AND TREATMENT OF DIFFERENTIATED THYROID CARCINOMA

JASNA MIHAILOVIĆ

Vojvodina Institute of Oncology, Sremska Kamenica, Serbia

BRANKO MARKOSKI

Faculty of Technical Sciences, University of Novi Sad, Novi Sad, Serbia

The current strategy of differentiated thyroid carcinoma (DTC) includes adequate initial treatment, which is defined as surgery (total or «near» total thyroidectomy) and radioactive iodine (131I) ablation of thyroid residues. 131I should be performed in all DTC patients post-surgically, except in low-risk patients (pTNM stage I). Despite many controversies regarding therapeutic dose of 131I, it has been recommended that the treatment with 3.7 GBq should be applied for thyroid residue ablation, 5.55 GBq for lymph node and/or lung involvement and 7.4 GBq for bone metastases. Due to regulatory limits, which differ from country to country, 131I therapy may require patient's hospitalisation or could be performed as outpatient treatment. In Serbia, the patient has to be hospitalised in isolated room until the emitted radiation flux falls below 0.40 GBq (10.8 mCi). Nowadays, 131I-whole body scintigraphy (WBS) with a diagnostic dose of 131I is usually not recommended before ablation. Post-ablation 131I-WBS is aimed at discovering possible undetected foci of uptake outside the thyroid bed. Besides other radionuclide agents, such as 111In octreotide, 99mTc-tetrofosfin, 99mTc-MIBI, and 99mTc(V)-DMSA, radioiodine-131 remains the best imaging tracer for localising the residue and/or recurrent disease.

1. Introduction

Thyroid nodules are clinically detectable in about 4–7% of the general population. The prevalence is even higher in the endemic regions with severe iodine deficiency. Most thyroid nodules are benign in nature, but thyroid carcinomas also appear.[1] The annual incidence of thyroid cancer rose for about 50% from 1973 to 1999. Due to the fact that thyroid cancer affected nearly 19000 people and caused about 12000 deaths in the United States in 1999, it was necessary to detect thyroid cancer in an early stage.[2]

2. Diagnosis and Treatment of Differentiated Thyroid Carcinoma

2.1. *Diagnosis of Differentiated Thyroid Carcinoma*

After the occurrence of thyroid ultrasound, the importance of thyroid scintigraphy in diagnostic evaluation of thyroid nodules has declined. When indicated, thyroid scintigraphy is performed with technetium (^{99m}Tc) or with radioactive iodine (^{131}I or ^{123}I), usually as a planar scintigraphy.[1] Some authors also suggested the value of the thyroid gland tomoscintigraphy in improving the sensitivity of thyroid imaging.[3] The role of thyroid scintigraphy in evaluation of a thyroid nodule is primarily to determine the functional status of the nodule to the surrounding thyroid parenchyma, i.e. to determine if the nodule is non-functioning ("cold"), functioning ("warm"), or hyper-functioning ("hot"). Cold nodules trap radioisotope less than the surrounding tissue, while warm or hot nodules trap as much or more tracers than the surrounding tissue.[4] Over 80% of all thyroid nodules are cold. The incidence of malignancy in cold nodules is 20–35%.[1,4] Due to the fact that standard imaging techniques, including scintigraphy and ultrasound, cannot reliably distinguish between benign and malignant thyroid tumours, there were several reports about radionuclide angiography of the thyroid nodes.[5,6] Thus, Mihailovic showed that radionuclide angiography is useful as a new additional imaging diagnostic tool in differentiation of cold thyroid nodes.[7] Contrary to that, the incidence of malignancy in warm or hot nodules is very low.[1,4,8]

Thyroid ultrasonography is becoming increasingly popular as the first-line diagnostic tool in evaluation of thyroid nodules, mostly without radionuclide imaging. Fine-needle aspiration cytology (FNAC) is guided by ultrasonography and has a specificity and sensitivity of 95% in the differential diagnosis of thyroid nodules.[1,9]

Therefore, in many cases patients undergo FNAC or even excision of thyroid nodules without previous radionuclide imaging. The role of nuclear medicine physician is usually in evaluation or consultation after thyroid carcinoma has been diagnosed.[4] Thyroid carcinomas include well-differentiated tumours which account for about 90% and non-differentiated tumours. Most differentiated thyroid carcinomas are papillary (70–80%), and follicular (20–30%). Non-differentiated carcinomas include medullar and anaplastic carcinomas, lymphoma and other less common tumours.[4]

2.2. *Management of Differentiated Thyroid Carcinoma*

Different guidelines were established to define the optional treatment for differentiated thyroid carcinoma (DTC), which has been controversial for decades.[10–14] The initial treatment of DTC implies an adequate initial treatment, i.e. surgery (total — TT or "near" total thyroidectomy — NTT) followed by radioactive iodine (^{131}I) ablation of normal and/or tumour residues in most cases.[15]

2.2.1. *Radioiodine therapy*

Radioiodine ablation of post-surgical thyroid residues should be performed as soon as possible after the NTT or TT, to ensure the best possible prognosis for DTC patients.[16] Thyroid residue ablation has been defined as the ^{131}I destruction of post-surgically residual macroscopically normal thyroid tissue.[2] In the United States, Beierwaltes started to perform the ^{131}I therapy in DTC patients in 1950.[17]

^{131}I ablation is usually not indicated in low-risk patients (pTNM stage I). On the contrary, it is highly recommended in patients with distant metastases, incomplete tumour resection and patients with complete excision of tumour but high risk of mortality and recurrence (pTNM stages II and III).[1] If residue ablation should be performed, the ^{131}I dose has to be identified. Maxon et al. proposed a patient-specific dosimetry method, which suggested that the selected dose should deliver at least 300 Gy to the thyroid residues for successful ablation.[18] On the other hand, some advocate a fixed dose for all DTC patients.[4] Whilst one favour low application in case of post-surgical discrete residues and the dose of ^{131}I ranging between 0.925 GBq[19] up to the value from 1.1 GBq to 1.85 GBq,[20–22] others favour a high application of 3.7 GBq,[23,24] especially in patients with significant residues.[15,25–27] According to the meta-analysis, both high and low activities are most likely to completely ablate the post-surgical thyroid residues.[25]

The total ablation is verified if there is no visible uptake in the thyroid bed, or if visible, fractional uptake is less than 0.1% on ^{131}I-whole body scintigraphy (WBS).[23] The beneficial effects of ^{131}I ablations were reported by many authors. Mazzaferri studied patients with larger tumours (>1.5 cm in diameter) and detected a recurrence rate of 16% in ^{131}I-treated group at 30 years, which was significantly less than 38% of the rate obtained in patients who were treated with thyroxin alone. Cancer-related deaths were detected in 3% of the ^{131}I-treated patients compared to 8% in patients treated without ^{131}I after 30 years.[2] A prospective multi-centre study has showed that ^{131}I reduces cancer-specific mortality and recurrence rates in high-risk DTC patients, with an additional improvement of disease-free survival.[28] Malesević *et al.* detected the survival of 91% after 18 years in DTC patients treated by surgery and ^{131}I therapy.[29]

Licensing of possession and discharge of ^{131}I for treating the patients varies from country to country. In Serbia, ^{131}I treatment requires hospitalisation of patients in a private room with a shower and a toilet to avoid radiation exposure to other patients, hospital personnel and the public. Patients cannot be discharged from hospital before the emitted radiation flux falls below 0.40 GBq.[30] Radioiodine therapy of DTC was performed in Serbia at the Institute of Oncology and Radiology in Belgrade in 1958 for the first time. The therapy was administered until 1990. After 1990, ^{131}I therapy is performed at the Institute of Oncology in Sremska Kamenica only. The radioiodine treatment has been performed at the Institute of Oncology in Sremska Kamenica since 1977, according to the Protocol for diagnosis, treatment and follow-up of DTC patients,[30] which was published in 1994.[31] Radioiodine therapy is safe and comfortable treatment and it rarely causes complications. Short-term complications are usually not serious and they are transient. Long-term complications are rare and may include gonad dysfunction, leukaemia (if cumulative doses of ^{131}I exceed 37 GBq) and pulmonary fibrosis.[32–35]

After the initial therapy, the follow-up of patients is aimed at maintaining an adequate thyroxin therapy and detecting the persistent or re-current disease. The follow-up should be routinely performed by serum thyroglobulin measurement, neck ultrasonography and ^{131}I-WBS (in cases of elevated thyroglobulin levels and positive ultrasound finding). Regarding the fact that recurrences are usually detected during the early years of follow-up, although they are also discovered later, the follow-up of DTC patients is necessary throughout the patient's life.[36] A local or regional recurrences occur in 5% to 20% of cases,[1,37] whereas distant metastases, usually in lungs and bones, in up to 10% of cases.[1]

Once it has been detected that a DTC patient has functioning metastases, the treatment dose must be selected. Most of the nuclear medicine physicians use «standard fixed-dose» ranging between 3.7 GBq and 7.4 GBq.[4] In the Institute of Oncology of Vojvodina in Sremska Kamenica the ^{131}I therapy in metastatic disease is performed according to the Protocol which recommends the application of 5.55 GBq in lymph and/or lung involvement and application of 7.4 GBq in bone metastases.[31] Stefanović et al have analysed DTC patients with distant metastases after the ^{131}I therapy. They showed that probability of cumulative survival was 62.8% after 5 years and 48.9% after 10 years.[38] They also found that treatment outcome was significantly influenced by localisation of metastases, their ability to accumulate ^{131}I, histological type of tumour and patient's age.[39] In another study, Stefanović *et al.* reported a higher frequency of distant metastases in patients with follicular carcinomas; bone metastases were more frequent in follicular carcinomas, while lung metastases occurred more often in papillary carcinomas.[40] Mihailovic *et al.* showed in their study that some predicting factors had significant impact on survival in DTC patients with distant metastases: age ($p = 0.0001$), histological type of tumour ($p = 0.0138$) and initial therapy ($p = 0.0351$), while gender had no impact. They concluded that longer survival of these patients could be achieved by adequate surgery followed by ^{131}I therapy.[41] Same authors detected that the absence of radioiodine uptake in distant metastases is a predicting factor which significantly shortens the survival of DTC patients with distant metastases.[42]

In metastatic disease, the repeated doses of ^{131}I may be required with each application ranging from 3.7 GBq to 10.1 GBq at 6-month interval between doses. There is no maximum limit to the cumulative dose of ^{131}I that can be applied to patients with metastatic disease, provided that individual doses do not exceed 2 Gy of the total body exposure. Most remissions are obtained with cumulative application of less than 22.2 GBq. Despite the fact that the risk of cancer and leukaemia rises with higher cumulative doses, there is no limit to the cumulative dose of ^{131}I in patients with distant metastases. If there is no uptake on post-therapy ^{131}I-WBS any further radioiodine therapy is useless.[1]

2.2.2. *Whole body scintigraphy*

The whole body imaging with ^{131}I is an imperative in evaluating the extent of thyroid carcinoma in the following cases: after the total or near-total thyroidectomy, during the regular follow-up of DTC patients if thyroglobulin serum level is elevated, in patients with detected metastases and during the check-up of patients following the ^{131}I therapy of metastases.[4]

Whilst the ^{131}I-WBS establishes the presence of residue thyroid tissue following surgery in DTC patients, the recent studies suggest that the tracer dose of ^{131}I may affect the efficacy of the subsequent radioiodine therapy. Standard doses for ^{131}I-WBS usually range from 74 MBq to 370 MBq and they allow identification of residual thyroid tissue and could also identify metastatic disease. Some believe that high application of ^{131}I can cause «stunning» effect.[43] Stunning effect occurs if the initial diagnostic dose affects the iodine trapping or organification mechanism so that uptake of subsequent therapeutic dose is impaired.[4] There are some data about differences between uptake of diagnostic and therapeutic doses.[44,45]

To overcome the problem with stunning, doses from 74 MBq to 185 MBq are recommended. This may reduce the stunning effect, but on the other hand, lower doses may reduce the sensitivity of imaging for detecting metastatic disease. Therefore, pre-therapeutic 131I-WBS is recommended to be performed with 123I,[46] which has sensitivity of 75% compared with ^{131}I if application of 74 MBq is used.[47] Others suggest a pre-therapeutic ^{99m}Tc scan to estimate the amount of the remaining normal thyroid tissue. However, it is still controversial to confirm a reduced uptake of the therapeutic dose as a consequence of the prior diagnostic dose rather than of the larger therapeutic dose.[4] A post-therapy ^{131}I scan should be performed in all patients, particularly in those who have not diagnostic ^{131}I-WBS (Fig. 1).

Following the successful ablation, the ^{131}I-WBS will show no abnormal foci of uptake in the absence of metastatic disease. Patients with metastatic disease, however, require regular ^{131}I-WBS between treatments to determine the efficacy of the ^{131}I therapy.[43]

Patients should be prepared prior to ^{131}I-WBS, and ^{131}I therapy by withdrawing thyroid hormone replacement, 2 weeks for triiodothyronine (T3) and 4 to 6 weeks for L-thyroxine (T4), in order to allow endogenous thyroid stimulating hormone (TSH) to increase to supra-normal levels, at least 30 mU/L. The measurements of thyroglobulin as a tumour marker in a routine follow-up is the most sensitive for detection of thyroid and/or metastatic tissue when it is performed during an adequate TSH stimulation. TSH stimulation may also be achieved by injecting recombinant human TSH (rhTSH), which stimulates thyroid tissue without requiring the discontinuation of thyroid hormone therapy.[4,33] Therefore, the use of rhTSH is indicated either for patients who undergo ^{131}I-WBS or radioiodine treatment, usually in those who cannot tolerate thyroxin withdrawal and who cannot generate endogenous TSH.[23] A diet low in iodine for at least 2 weeks before ^{131}I-WBS is also suggested to augment the ^{131}I uptake and retention, as shown in Table 1 and Table 2.[4]

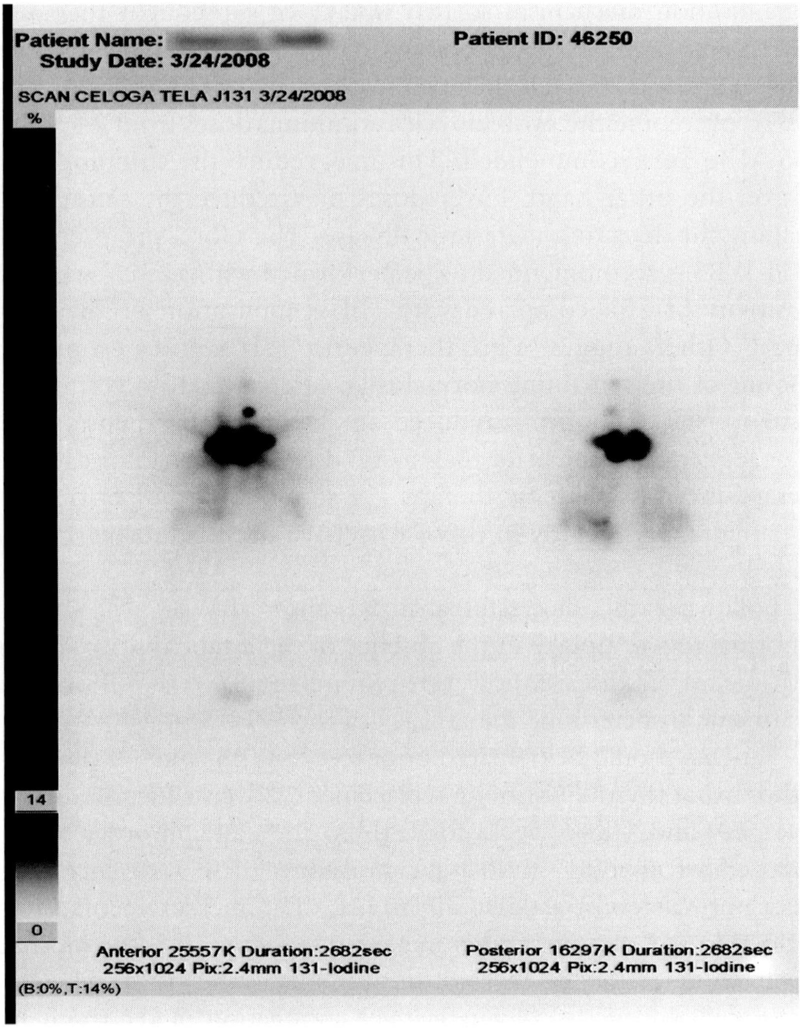

Figure 1. Post-therapy ^{131}I scan shows large postoperative thyroid remnants; regional metastases in cervical lymph nodes; and distant metastases in upper mediastinal lymph nodes and lungs.

2.2.3. *Follow-up*

The diagnostic WBS should be performed in all DTC patients who underwent the ^{131}I therapy six months following the radioiodine treatment. In some cases, ^{131}I-WBS will be found to be negative with

Table 1. Low iodine diet-forbidden foods

Forbidden foods
1. Iodised salt, sea sault (specifically salted nuts, crackers, and pretzels)
2. Milk or diary products (e.g. cheese, chocolate, ice cream, yogurt)
3. Eggs
4. Seafood (e.g. shellfish, kelp, seaweed)
5. Foods that contain the additives carrageenan, algin, alginate, agar-agar
6. Breads made with iodate dough conditioners
7. Foods and medications containing red food dyes (commercial canned tomato sauces, cereals, candies, vitamins)
8. Mineral supplements
9. Iodinated medications (radiographic contrast material)
10. Cured and corned foods (e.g. ham, lox, corned beef, sauerkraut)
11. Soy products (e.g. soy sauce, soy milk)

Table 2. Low iodine diet-allowed foods

Allowed foods
1. Red meat, pork, veal, chicken
2. Potatoes and vegetables (cooked without iodized salt)
3. Pasta and rice
4. Fresh vegetables includes tomatoes and sauces

elevated thyroglobulin levels, due to several reasons. Patients with Hürthle cell carcinomas are iodine avid in only 10% of cases and there is also the possibility of re-differentiation of papillary and follicular thyroid carcinomas. According to some data, about 30% of patients will show successful re-differentiation and restoration of tumour iodine uptake capabilities with retinoid acid.[43]

There are several alternative radionuclide agents which could be used to detect localisation of the re-current disease in ^{131}I scan negative patients. Some published that the sensitivities of ^{201}Tl and ^{99m}Tc tetrophosphine are comparable to that of ^{131}I for detecting distant metastases (0.85; 0.85; 0.78) beside the fact that ^{131}I is more sensitive than the other two for detecting post-surgical thyroid residues.[48] ^{99m}Tc sestamibi (MIBI) is clinically more useful than ^{201}Tl in detecting lung, lymph node and bone metastases in DTC, as image quality is better.[49] Both MIBI and ^{201}Tl have high specificity and positive predictive value for residual thyroid carcinoma in patients with negative ^{131}I-WBS who have an increased risk of recurrence after radioiodine therapy.[50] It has been reported that ^{99m}Tc tetrophosphine has a high sensitivity in detecting metastases and recurrences in DTC patients.[51] Recently, some preliminary reports suggested the role for ^{111}In octreotide in detecting recurrence in DTC patients with negative ^{131}I-WBS.[52] ^{18}FDG-PET has proved to be useful in detecting cervical lymph nodes, mediastinal metastases and neoplastic tissue in thorax and skeleton. Usually, the sensitivity of ^{18}FDG-PET studies of 70% to 90% has been reported in DTC patients with elevated thyroglobulin levels and negative ^{131}I scan.[53,54] Actually, the main value of ^{18}FDG-PET in the management of DTC patients, lies in its ability to detect metastatic sites in patients with elevated thyroglobulin levels and negative ^{131}I scans. A multi-centre study comparing the sensitivity of ^{18}FDG-PET, ^{201}Tl, and ^{99m}Tc-MIBI in ^{131}I scan negative disease, showed significantly higher sensitivity (90%) of that compared to other scanning agents.[55]

There is also another agent, ^{99m}Tc (V)-DMSA, which can be used in Hürthle cell carcinomas (variants of follicular thyroid cancer) and in poorly differentiated and highly aggressive cancers called insular carcinomas. Poor imaging results have been obtained with most radionuclides including ^{131}I, ^{111}In octreotide, ^{99m}Tc-MIBI, and ^{99m}Tc (V)-DMSA.[56–58]

3. Conclusion

Radionuclide isotopes in nuclear medicine have an established role in the management of patients with differentiated thyroid carcinoma. Radioiodine-131 remains the routine radionuclide in imaging technique for detection of thyroid residues and the re-current disease. Others are also available, such as ^{111}In-octreotide, ^{99m}Tc-tetrophosphine, ^{99m}Tc-MIBI and ^{99m}Tc (V)-DMSA, but with limited diagnostic value. In the field of therapy, ^{131}I still remains the best choice as the most sensitive radionuclide in treatment of DTC patients.

References

1. M. Schlumberger and F. Pacini, Thyroid Tumours, Editions Nucleon, Paris (2006).
2. E. Mazzaferri and R.T. Kloos, *J. Clin. Endocrinol. Metab.* **86**, 1447 (2001).
3. Lj. Stefanović, M. Malešević, E. Kamenicki, J. Mihailović and D. Srbovan, *Radiol. Arch. Srb.* **5**, 1009 (1996).
4. S.J. Goldsmith, *Thyroid Carcinoma, Nuclear Oncology: Diagnosis & Therapy*, **197** (Lippincott Williams & Wilkins, 2001).
5. J. Mihailović, *Eur. J. Nucl. Med.* **18**, 583 (1991).
6. J. Mihailović, *J. Endocrinol. Investigat.* **15**, 53 (1992).
7. J. Mihailović, *Med. Pregl.* **45**, 101 (1992).
8. M. Malesevic, Lj. Djilas, J. Mihailović, R. Cvijanovic, B. Hadzic and J. Bogdanovic, *Med. Pregl.* **44**, 48 (1991).
9. Lj. Stefanović, M. Malesevic, J. Mihailovic, O. Adzic, N. Sljapic and M. Berger, *Med. Pregl.* **44**, 32 (1991).
10. C.J.H. Van de Velde, J.F. Hamming, B.M. Goslings, L.J.D.M. Schelfhout, O.H. Clark and S. Smeds, *Eur. J. Cancer. Clin. Oncol.* **24**, 287 (1988).
11. P.A. Singer, D.S. Cooper, G.H. Daniels, P.W. Ladenson, F.S. Greenspan and E.G. Levy, *Arch. Intern. Med.* **156**, 2165 (1996).
12. F. Pacini, M. Schlumberger, H. Dralle, R. Elisei, J.W.A. Smit and W. Wiersinga, *Eur. J. Endocrinol.* **154**, 787 (2006).
13. The British Thyroid Association and the Royal College of Physicians, www.british-thyroid-association.org/guidelines.htm (2006).
14. D.S. Cooper, G.M. Doherty, B.R. Haugen, R.T. Kloos, S.L. Lee and S.J. Mandel, *Thyroid* **16**, 109 (2006).
15. P.W.S. Rosario, A.L. Barroso, L.L. Rezende, E.L. Padrao, T.A. Fagundes and J.S. Reis, *Ann. Nucl. Med.* **19**, 247 (2005).
16. F.A. Verburg, B. de Keizer, C.J. Lips, P.M. Zelissen and J.M. de Klerk, *Eur. J. Endocrinol.* **152**, 33 (2005).
17. W.H. Beierwaltes, R.H. Nishitama, N.W. Thompson, J.E. Copp and A. *J. Nucl. Med.* **23**, 561 (1982).
18. H.R. Maxon, E.E. Englaro, S.R. Thomas, V.S. Hertzberg, J.D. Hinnefeld and L.S. Chen, *J. Nucl. Med.* **33**, 1132 (1992).
19. C.S. Bal, A. Kumar and G.S Pant, *J. Clin. Endocrinol. Metab.* **89**, 1666 (2004).

20. K. Johansen, N.J. Woodhouse and O. Odugbesan, *J. Nucl. Med.* **132**, 252 (1991).
21. P.W. Rosario, J.S. Reis, A.L. Barroso, L.L. Rezende, E.L. Padrao and T.A. *Nucl. Med. Commun.* **25**, 1077 (2004).
22. C. Bal, A.K. Padhy, S. Jana, G.S. Pant and A.K. Basu, *Cancer* 77, 2574 (1996).
23. F. Pacini, P.W. Ladenson, M. Schlumberger, A. Driedger, M. Luster and R.T. Kloos, *J. Clin. Endocrinol. Metab.* **91**, 926 (2006).
24. W.H. Beierwaltes, R. Rabbani, C. Dmuchowski, R.V. Lloyd, P. Eyre and S. Mallette, *J. Nucl. Med.* **25**, 1287 (1984).
25. S.A. Doi and N.J. Woodhouse, *Clin. Endocrinol.* **52**, 765 (2000).
26. D.E. Roos and J.G. Smith, *Int. J. Rad. Oncol. Bio. Phys.* **44**, 493 (1999).
27. M. Gawkowska-Suwinska, M. Turska, J. Roskosz, Z. Puch, B. Jurecka-Tuleja and D. Handkiewicz-Junak, *Wiad. Lek.* **54**, 278 (2001).
28. T. Taylor, B. Specker, J. Robbins, M. Sperling, M. Ho and K. Ain, *Ann. Intern. Med.* **129**, 622 (1998).
29. M. Malesevic, J. Vojicic, Lj. Stefanović, J. Mihailović, B. Guduric and J Bogdanovic, Survival of differentiated thyroid carcinoma patients treated by surgery and by ^{131}I from 1977 to 1998, Messenger of Institute for thyroid gland and metabolism "Zlatibor", Zlatibor, 67 (2002).
30. J. Mihailović, *Arch. Oncol.* **14**, 45 (2006).
31. Lj. Stefanović, B. Guduric, N. Sljapic, J. Babic, Lj. Miljkovic and M. Malesevic, *Arch. Oncol.* **2**, 101 (1994).
32. M. Dietlin, D. Moka and H. Schicha, *Radioiodine Therapy for Thyroid Cancer*, **95** (Springer-Verlag, 2005).
33. R. Cavalieri, *Nuclear Medicine, Thyroid Cancer: Diagnosis and Treatment*, 161 (Quality Medical Publishing, 2000).
34. M. Malesević and J. Mihailović, Therapy of differentiated thyroid carcinoma (DTC) by radioioiodine 131-I, *UONS Symposium: Carcinoma of the thyroid gland. Belgrade, Serbia,* 1 (2004).
35. J. Mihailović, M. Malesevic and Lj. Stefanović, I therapy in the treatment of differentiated thyroid carcinoma, *5th Symposium of Serbian Assotiation of oncology surgeons,* Kopaonik, Serbia, Revial lectures, 23 (2006).
36. M. Malesevic, J. Mihailović, J. Vojicic and S. Popadic, *Acta. Chirurgica. Iugoslavica.* **L(3)**, 177 (2003).
37. Lj. Stefanović, J. Mihailović and M. Malesevic, *Eur. J. Nucl. Medic.* **34**, 456 (2007).

38. Lj. Stefanović, J. Vojicic, M. Malesevic, J Mihailović and D Srbovan, Influence of distant metastases to the survival of differentiated thyroid carcinoma patients, *The 2nd congress dedicated to the thyroid gland-Zlatibor 2000, Serbian Academy of Sciences and Art*, Medical Faculty Belgrade, 173 (2001).
39. Lj. Stefanović, J. Vojicic, M. Malesevic, J. Mihailović and D. Srbovan, *J. BUON* **6**, 263 (2006).
40. Lj. Stefanović, M. Malesevic, J. Mihailovic, D. Srbovan, and J. Vojicic, *Medicus* **2**, 31 (2001).
41. J. Mihailović, Lj. Stefanović and M. Malesevic, *Cancer Bioth. Radioph,* **22**, 250 (2007).
42. J. Mihailovic, Lj. Stefanović and M. Malesevic, *Eur. J. Nucl. Medic.* **234**, 511 (2007).
43. S.E.M. Clarke, *Diagnostic Nuclear Medicine Investigations in the Management of Thyroid Cancer, Practical Management of Thyroid Cancer, A Multidisciplinary Approach*, 341 (Springer-Verlag, 2006).
44. H. Park, Stunned thyroid after high-dose I-131 imaging, *Clin. Nucl. Med.* **17**, 501 (1992).
45. R.K. Jeevanram, D.H. Shah, S.M. Sharma and R.D. Ganatra, Influence of initial large dose on subsequent uptake of therapeutic radioiodine in thyroid cancer patients, *Nucl. Med. Biol.* **13**, 277 (1986).
46. H.M. Park, O.W. Perkins, J.W. Edmonson, R.B. Schnute and A. Manatunga, *Thyroid* **4**, 49 (1994).
47. E.J. Murphy, F. Almeida, R. Lull and J.E. Kalinyak. *J. Invest. Med.* **70A** (2000).
48. S. Unal, Y. Menda, I. Adalet, H. Boztepe, N. Ozbey and F. Alagol, *J. Nucl. Med.* **39**, 1897 (1998).
49. S. Miyamoto, K. Kasagi, T. Misaki, M.S. Alam and J. Konshi, *J. Nucl. Med.* **38**, 352 (1977).
50. J.E. Seabold, N. Gurll, M.E. Schurrer, R. Aktay and P.T. Korchner, *J. Nucl. Med.* **40**, 1434 (1999).
51. H.J. Gallowitsch, P. Mikosch, E. Kresnik, O. Unterweger, I. Gomez and P. Lind, *J. Nucl. Med.* **39**, 870 (1998).
52. E. Baudin, M. Schlumberger, J. Lumbroso, J.P. Travagli, B. Caillou and C. Parmentier, *J. Clin. Endocrinol. Metab.* **81**, 2541 (1996).
53. J.K. Chung, Y. So, J.S. Lee, C.W. Choi, S.M. Lim and D.O. Lee, *J. Nucl. Med.* **40**, 986 (1999).

54. W. Wang, *J. Clin. Endocrinol. Metab.* **84**, 2291 (1999).
55. F. Grunwald, T. Kalicke, U. Feine, R. Lietzenmayer, K. Scheidhauer and M. Dietlein, *Eur. J. Nucl. Med.* **26**, 1547 (1999).
56. F.R. Tenenbaum, J. Lumbroso, M. Schlumberger, B. Caillou, P. Fragu and C. Parmentier, *J. Nucl. Med.* **36**, 807 (1999).
57. I.T. Zak, J.E. Seabold and N.J. Gurll. *Clin. Nucl. Med.* **20**, 31 (1995).
58. T.C. Yen, K.L. King, A.H. Yang, R.S. Liu and S.H. Yeh, *J. Nucl. Med.* **37**, 78 (1996).

CHAPTER 5

RESULTS OF DENSITOMETRIC MEASUREMENT BY HOLOGIC–QDR–2000 IN THE PERIOD FROM 2005 TO 2007

OLGICA MILETIĆ

Railway Health Care Institute, Nis, Serbia

TATJANA JOVANOVIĆ

Faculty of Medicine, University of Nis, Nis, Serbia

Osteoporosis is a bone disease, marked by low bone mass and micro-architectural deterioration that leads to an increased susceptibility to fracture. Fractures are the single most important clinical consequence of osteoporosis and represent the major health problem in the elderly. Bone density, a measure of bone mass, can be measured non-invasively by using widely available densitometric techniques, and is the main focus of risk assessment and therapy. The main objective of this study is to systematise and show the results in the period of two years. Measurements were taken at the Hologic–QDR–2000 device. Also, this systematisation should be used as an addition to the available data base at the Institute of Rheumatology and Physical Medicine.

1. Introduction

Osteoporosis, a disease defined by low bone mass and deterioration in bone structure, leads to the occurrence of fractures after relatively low levels of trauma. The most common sites for osteoporotic fractures are the spine, hip and wrist. However, all fractures that occur from low trauma are considered osteoporotic fractures. The most common osteoporotic fractures occur in the vertebrae. Vertebral fractures can be completely asymptomatic and fewer than 15% of these fractures ever come to medical attention.

This fact does not lessen the importance of vertebral fractures, as such events are the strongest risk factor for additional fractures.[1–4] Risk factors for osteoporosis include the female gender, increased age, estrogen deficiency, white race, low weight and body mass index (BMI), family history of osteoporosis, smoking etc.

In 1994, the World Health Organization (WHO) provided criteria for diagnosis of a normal bone mass, low bone mass or osteopenia, and osteoporosis (Table 1). The WHO criteria for osteoporosis were based on comparisons to the peak adult bone mass (PABM) from a population of Caucasian post-menopausal women. Using 2.5 standard deviations below PABM as the definition of osteoporosis (T-score of — 2.5), a 30% prevalence of osteoporosis was found among older post-menopausal Caucasian women.

Table 1. World Health Organization (WHO) — Criteria for diagnosis of a normal bone mass, low bone mass or osteopenia and osteoporosis*

Normal	**BMC or BMD not more than 1 standard deviation below the peak adult bone mass**
Osteopenia	BMC or BMD that lies between 1 and 2.5 standard deviations below the peak adult bone mass. T-score between –1 and –2.5
Osteoporosis	BMC or BMD value more than 2.5 standard deviations below the peak adult bone mass. T-score ≤ –2.5
Severe Osteoporosis	BMC or BMD value more than 2.5 standard deviations below the peak adult bone mass and the presence of one or more fragility fractures. T-score ≤ –2.5 plus fragility fracture

*World Health Organization criteria for diagnosis of osteoporosis based on bone mineral content (BMC) or bone mineral density (BMD) measurements. These criteria can be applied to either the central or peripheral skeletal measurement sites.[3–5]

Although the WHO criteria identify individuals with high fracture risks, the risk of fracture is, in fact, a continuous gradient that increases with decreasing bone mineral density (BMD) levels.[3–5]

2. Measurements and Results

The results of measurements have been collected from the records taken at the HOLOGIC–QDR–2000 which is used in the Railway Health Care Institute to diagnose the condition of a patient. The Hologic QDR–2000 X-ray Bone Densitometer images and measures the BMC and BMD of selected areas of the body, or of the entire skeleton. It does so by using X-rays of two different energy levels. This dual-energy scheme allows soft tissue within the selected area to be subtracted out, leaving only bone to be imaged and measured.

The patient entrance dose is less than 1,29 μC/kg, approximately one–tenth the exposure of a standard chest X-ray. Leakage radiation at one meter is less than 14,33 pC/kgs, or approximately one–thousandth of the limit specified in 21 CFR 1020.30 (k). No additional shielding is necessary for a patient, operator or room, and the QDR–2000 can be placed in any convenient non-shielded examination room.

The QDR–2000 employs a patented Automatic Internal Reference System, which continuously calibrates the machine to eliminate the effects of variations in temperature, tube flux, etc. No daily calibration is required, though Hologic does recommended the daily scanning of a quality control phantom to provide assurance that the system is functioning correctly, and to assist in detection of any long-term drift.

Out of the total number of screening records in the period from 2005 till 2007, the results taken from 200 patients of different age (45–55, 55–65 and 65–75) have been analysed.

T-score, height and weight of the patients were used as the referential values. On the basis of the taken records for height (H) and weight (W), BMI values for each individual patient were calculated by using the formula:

$$BMI = \frac{W}{H^2}, \left[\frac{\text{kg}}{\text{m}^2}\right] \tag{1}$$

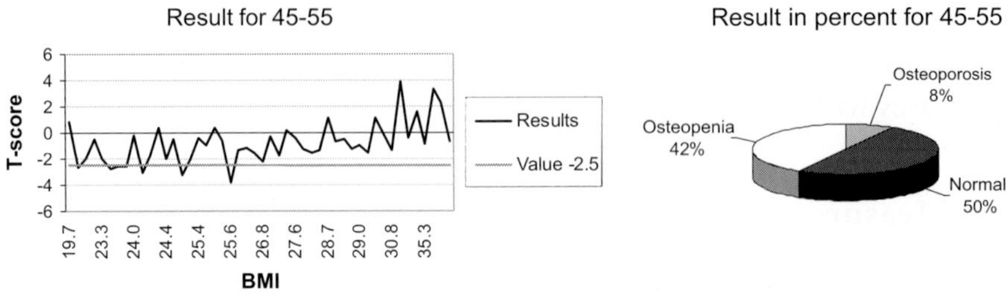

(a) (b)

Figure 1. Result for 45–55.

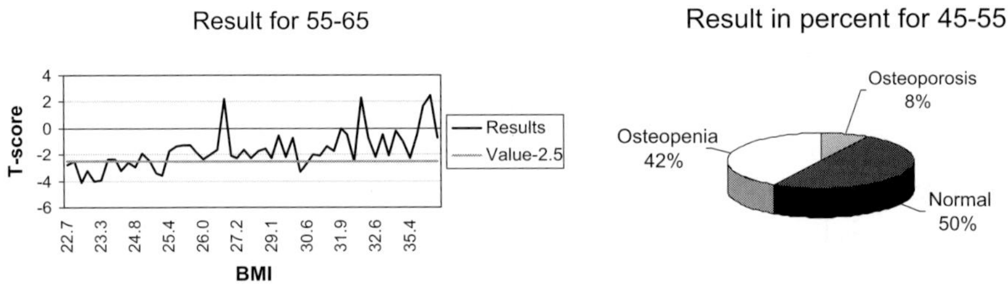

(a) (b)

Figure 2. Result for 55–65.

The analysed results of the T-score dependency on BMI are shown in Figure 1(a) for the group of patients aged 45–55, Figure 2(a) for 55–65 and Figure 3(a) for the group of patients aged 65–75. The horizontal line denotes the T-score value of −2.5.

The results show that the conclusions so far confirm that the low value of BMI represents one of the most significant risk factors of osteoporosis. The largest number of patients diagnosed for osteoporosis in all three age groups has low BMI, as shown in the graphics. Some variations were recorded in the age group from 45 to 55, in patients with some previous diagnosis (early surgical or natural menopauses).

Also, some variations were recorded in the age group from 65 to 75, due to an old age.

Figures 1(b), 2(b) and 3(b) show the inter-dependency between the T-score and age. The T-score values above −1 are highlighted in grey

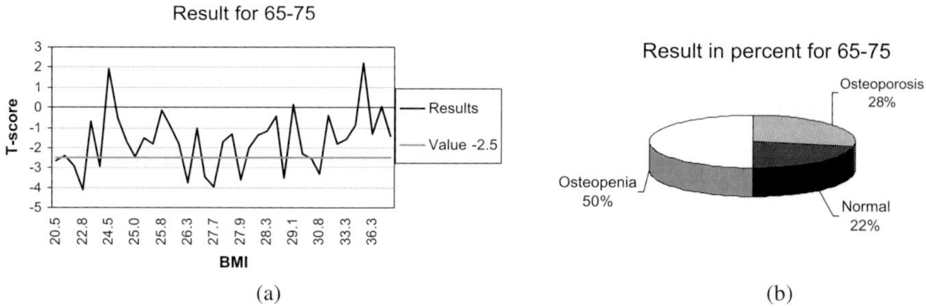

Figure 3. Result for 65–75.

colour, values from −1 to −2.5 in white and values below −2.5 are shown in dark colour.

The picture shows that the biggest number of healthy persons is in age group from 45 to 55, whereas the biggest number of patients with osteoporosis belongs to the age group from 65 to 75.

3. Conclusion

DEXA (dual-energy-X-ray absorptiometry) bone densitometry is most often used to diagnose osteoporosis, a condition that affects women after menopause but may also be found in men. Osteoporosis involves a gradual loss of calcium causing the bones to become thinner, more fragile and more likely to break. Bone density scanning, or bone densitometry, is an enhanced form of X-ray technology that is used to measure bone loss. Special care is taken during X-ray examinations to use the lowest radiation dose possible while producing the best images for evaluation. National and international radiology protection councils continually review and update the technique standards used by radiology professionals. DEXA is today's established standard for measuring bone mineral density (BMD). Osteoporosis is the most frequent bone disease and as such represents one of the most significant problems in the world, as well as in Serbia. On the basis of our results, it can be concluded that it is of outmost importance, unless some other risk factor is diagnosed, that the first screening should be undertaken at the age of 50 to prevent osteopenia, especially osteoporosis.

References

1. NIH Consensus Development Panel. Osteoporosis prevention, diagnosis, and therapy, *JAMA 2001*, 785 (2001).
2. M.C. Nevitt, B. Ettinger and D.M. Black, *Ann. Intern. Med.* **128**, 793 (1998).
3. Assessment of fracture risk and its application to screening for postmenopausal osteoporosis. Report of a WHO Study Group. *World Health. Organ. Techn. Rep. Ser.* **843**, 1 (1994).
4. P.D. Miller, S.L. Bonnick, Clinical application of bone densitometry, Primer on the Metabolic Bone Disease and Disorders of Mineral Metabolism, Lippincott-Raven, 152 (1999).
5. J.A. Kanis, *Osteoporos.* Int. 7, S108 (1977).

CHAPTER 6

TRAVELLER'S THROMBOSIS IN THE CITY OF TIMISOARA

IULIAN-OCTAVIAN AVRAM

City Hospital Timisoara, Timisoara, Romania

MIHAELA FLAVIA AVRAM

County Hospital Timisoara, Timisoara, Romania

The objective of our study was to determine the incidence and risk factors of traveller's thrombosis in Timisoara, Romania. During the period from January 2002 to January 2006 we studied 217 patients with suspected DVT, of which 153 were confirmed. Our study group consisted of 29 patients with confirmed DVT, with history of recent long-lasting travels by plane, car or by bus. 58.62% of patients were female patients. Additional risk factors were present in 28 (96.55%) patients from the study group. Thrombosis occurred mainly after travelling by bus (14 patients — 48.27%), by plane (9 patients — 31.03%) or by car (6 patients — 20.69%). No case of thrombosis after travelling by train was recorded. In 62.06 % of cases, the journey lasted for at least 12 hours, with the mean duration of 16.3 hours. Although the onset of clinical signs occurred after 12–24 hours (mean 14.3 hours), the mean interval that elapsed between travel and hospital admission was significantly higher, with a mean value of 5.1 (0–25) days, due lack of information among patients. The risk of DVT is increased by all types of travel. We found that DVT occurs more frequently after travelling by bus (48.27%) than travelling by plane or car. All patients should be informed about clinical signs of DVT, in order to avoid a long delay between onset of clinical signs and commencement of proper therapy.

1. Introduction

The freedom of travelling, wealth and economic progress determine an increase in number of travellers, but also in travel distance and duration, by plane or by bus. The death of a young woman right at the London Airport, on September 29th, 2000, shortly after a transcontinental flight, highlighted the risk of deep venous thrombosis (DVT) and pulmonary emboli (PE) following prolonged immobilisation, thus inducing a real hysteria in mass-media.

The first report on deep vein thrombosis (DVT) following prolonged air travel was published in 1954 by Homans, who described two cases of thrombosis occurring after flights from the USA to Venezuela.[1]

Homans' original report included three other patients with DVT that occurred after a prolonged sitting during travelling by car (two patients) and after attending a long performance at the theatre.

The term "economy class syndrome" was introduced in 1977 by Symington and Stack who reported a series of eight patients with venous thromboembolism (VTE) occurring shortly after flying economy class.[2] This term was considered inaccurate, as several studies noticed the association between long journeys involving prolonged immobilisation and risk of thrombosis after travelling by bus, car and train but also after travelling in a business-class.[3]

Since then, a substantial number of cases have been recorded. These reports highlight the important question as to whether travel is associated with venous thromboembolism or not, but they do not provide a concrete answer.

The consensus at a meeting of experts held in March 2001 and organised by the World Health Organization (WHO) to review the evidence was that a link "probably" exists, although the incidence is low and mainly involves passengers with additional risk factors for venous thromboembolism, but that the risk is not quantifiable because of the lack of adequate data.[4]

The term "economy class syndrome" was replaced by the term "traveller's thrombosis", considered to be a more appropriate term.[5,6]

2. Material and Methods

A comprehensive study of the patients hospitalised with a presumptive diagnosis of deep vein thrombosis (DVT) or pulmonary embolism (PE) was undertaken. We prospectively determined the risk factors for thromboembolism in a cohort of 217 consecutive inpatients suspected with DVT or PE attending the First Surgical Clinic Timisoara, two freestanding surgical units, and the Cardiology Clinic Timisoara, from January 2002 to January 2006. Referrals originated mostly from family doctors, but also from several ambulance surgery units. Medical history was reviewed, using a standardised form, focusing on questions regarding travel, risk factors like hormone therapy, malignancy, surgery, immobilisation, pregnancy, obesity and previous VTE. The details were recorded of journeys longer than 4 hours in the previous 30 days. The diagnosis of DVT was confirmed using duplex ultrasonography or phlebography in selected cases. The diagnosis of PE was confirmed by chest X-ray or ventilation–perfusion lung scanning in 9 cases.

Subjects were excluded if the diagnosis of DVT was not confirmed by one of the above criteria, if there was an alternative diagnosis or if the case represented the readmission for the same VTE event, occurring within 4 months from the first admission.

Thrombophilia tests were not routinely undertaken in these patients and for this reason could not be included in the analysis. Patients were periodically followed-up, for 3–12 months — mean period of 4.34 months. Statistical analysis of collected data was performed using GraphPad Instat, version 3.05.

3. Results

In the period from January 2002 to June 2006, we recorded 217 cases with suspected DVT or PE. Duplex ultrasonography confirmed 122 cases of DVT, 31 cases were confirmed by phlebography; 64 cases were excluded from the study, because the diagnosis of DVT was not confirmed. The mean age of patients with DVT was 57.6 years, and 82 (53.59%) of them were women.

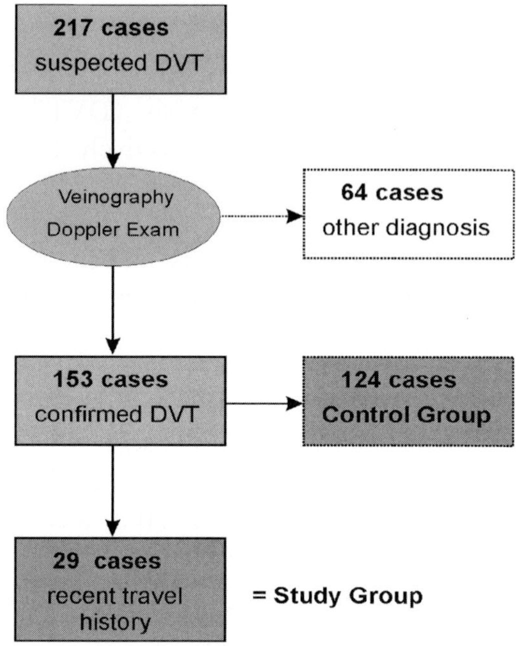

Figure 1. Algorithm for diagnosis of DVT and recent travel.

Recent history of air travel was recorded in 29 patients out of 153 cases confirmed with DVT (18.95%), see Figure 1, with travel time longer than 4 hours in the previous 30 days, with 9 (31.03%) of these having a history of plane travel exceeding 8 hours. The control group included 124 patients without travel history.

The main characteristics of the two groups were compared, but found no statistically significant difference (Table 1). Mean age of patients with travel-related thrombosis was 53.3, with a higher incidence among female patients — 17 (58.62%), see Figure 2.

Most patients reported progressive swelling of the leg, with onset in the first 7 days after a long-distance journey (Table 2). Although the onset of clinical signs occurred after 12–24 hours (mean 14.3 hours), see Figure 3, the mean interval between travel and hospital admission was significantly higher, with a mean value of 5.1 (0–25) days, see Figure 4.

Table 1. Clinical characteristics of study group compared to control group

		Study Group (29 P)	Control group (124 P)	P
Mean age		53,3 years	57,6 years	0.2571
Sex ratio	Male	12 (41.38%)	58 (46.78%)	0.7505
	Female	17(58.62%)	66 (53.22%)	
BMI	<25 kg/m^2	5 (17.24%)	31 (25%)	0.6825
	25–35 kg/m^2	20 (68.97%)	59 (47.58%)	
	>35 kg/m^2	4 (13.79%)	34 (27.42%)	
Diagnosis	DVT	27 (93.11%)	118 (92.74%)	0.7470
	DVT + PE	2 (6.89%)	9 (7.26%)	
Contraceptive use		2 (6.89%)	7 (5.64%)	
		1 (3.45%)	8 (6.45%)	
No risk factors				

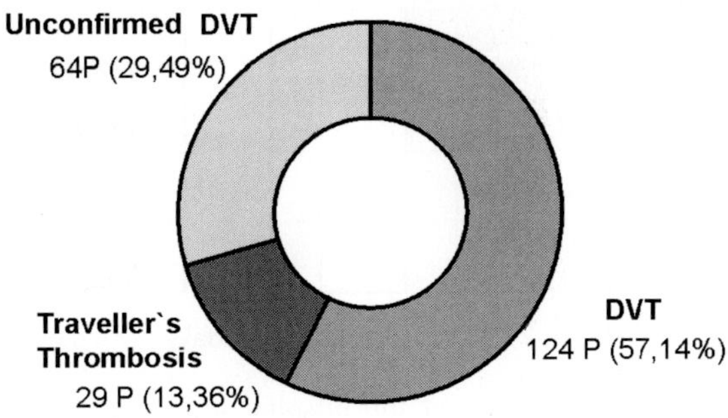

Figure 2. Patients with suspected DVT–clinical findings.

Table 2. Clinical signs recorded in the study group

Clinical signs	No. PTS	%
Oedema	26	89.65
Pain or tenderness in the leg	24	82.75
Feeling of increased warmth	15	51.72
Cyanosis of lower limb	9	31.03
Functional impairment	7	24.13
Shortness of breath	2	6.89

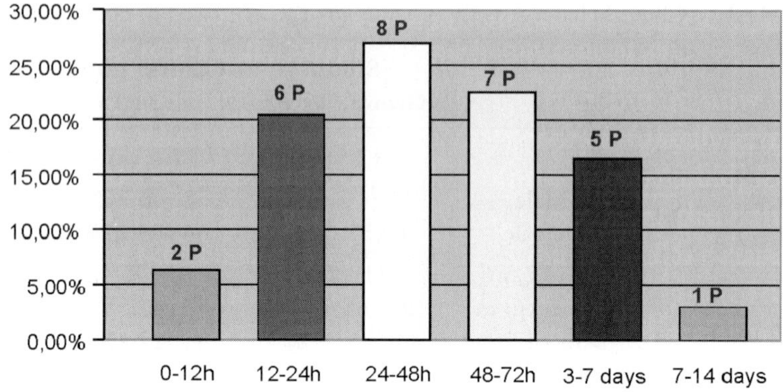

Figure 3. Interval between travel and onset of symptomatology in the study group.

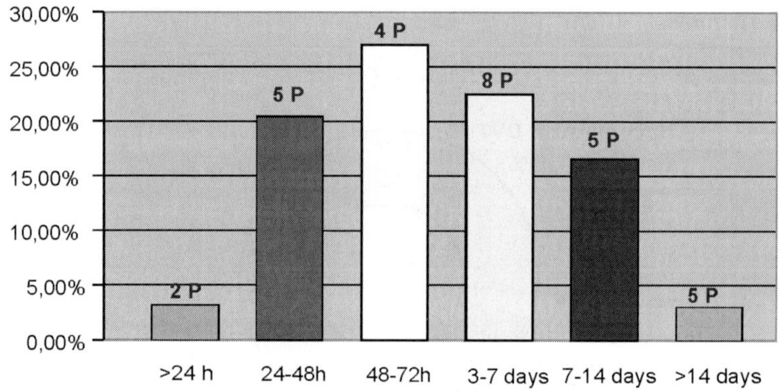

Figure 4. Interval between travel and hospital admission in the study group.

We noticed a higher incidence in patients aged 40–60 years compared to other age groups, with the mean age of 53.3 years, see Figure 5.

Almost 50% of DVT occurred after a long-distance travelling by bus (48.27%), followed by plane travel (31.3%) and car travel (20.68); no cases of DVT occurred after travelling by train, see Figure 6.

In 65.51% of cases, the travel duration was of at least 12 hours, with the mean duration of 16.3 hours (Table 3).

Risk factors were identified in 28 out of 29 (96.55%) patients with traveller's thrombosis (Table 2). One young male patient without

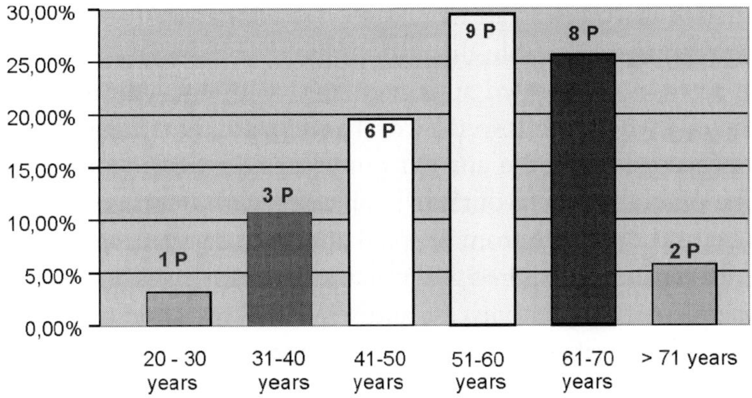

Figure 5. Distribution of cases by age groups.

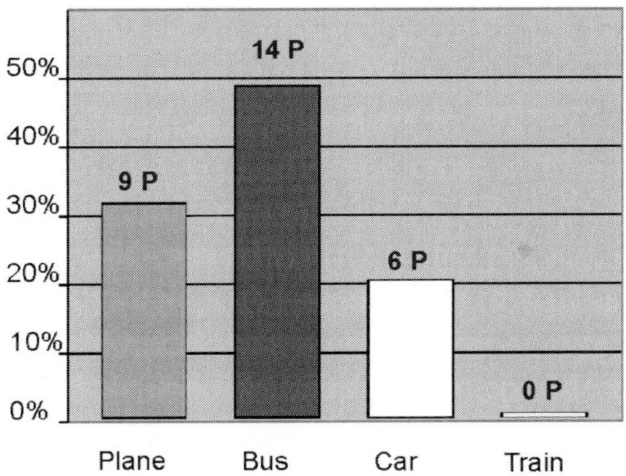

Figure 6. Distribution of cases by means of travel.

Table 3. Duration of travel

Duration of travel	No. PTS	%
4–6 h	3	10.34
6–12 h	6	20.68
12–24 h	14	48.27
>24 h	5	17.24
Not recorded	1	3.45

known risk factors was addressed to the Clinic of Haematology for further testing of possible thrombophilia (Table 4).

In all cases we applied anticoagulant treatment with unfractioned heparin or LMWH, followed by oral anticoagulant treatment, for 3–6 months according to the current guidelines for treatment of DVT.

The outcome was favourable in all cases, with improvement of clinical signs and discharge from hospital after 5–9 days (mean value of 6.7 days). Two patients required additional surgery for associated diseases — right hemicolectomy in one case and sequential resection of small bowel due to intestinal infarction in another case, see Figure 7.

Table 4. Risk factors in the study group

Risk factor	No. PTS	%
Chronic venous insufficiency	12	41.37
Congestive heart failure	5	17.24
Obesity	5	17.24
Family history	4	13.79
Previous VTE	3	10.34
Oral contraceptive pill	2	6.89
Hormone replacement therapy	1	3.45
Malignancy	1	3.45
Recent trauma	1	3.45
Recent surgery	0	0
Pregnancy	0	0
No known risk factors	1	3.45

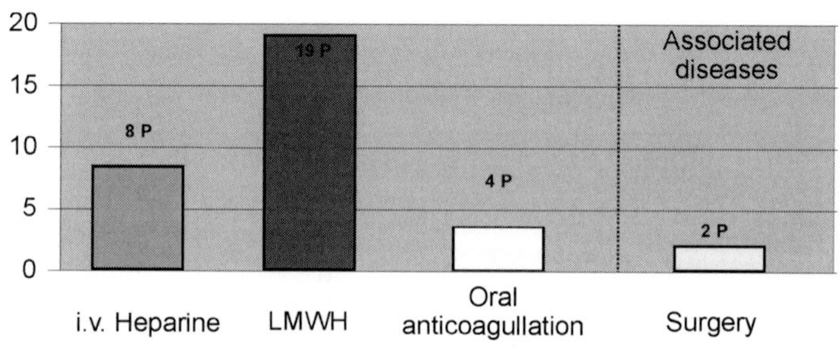

Figure 7. Treatment applied.

4. Discussion

The significance of travel-associated venous thrombosis remains unknown due to poor understanding of its natural history. It is well-established that inherited and acquired factors increase the risk of travel related thrombosis. The risk is particularly high in long-haul travellers over 40 years of age with co-morbid conditions such as cancer, congestive heart failure, recent surgery or previous VTE or stroke.[7] The risk is also high in passengers with varicose veins, obesity or recent trauma.[8] Patients with thrombophilia, pregnant women or women on oral contraceptives or hormone replacement therapy are also at an increased risk.[9]

Martinelli *et al.*[7] analysed 210 patients with DVT and found a 16-fold increased risk in patients who travelled by air and had some form of thrombophilia, as well as a 14-fold higher risk in women who flew and used oral contraceptives.

Prophylaxis of travellers' thrombosis is of the utmost importance. Passengers should be informed about potential risks and should be instructed to exercise regularly, to perform short walks and to avoid dehydration and alcohol intake.

Patients should also be informed about symptoms of DVT in order to avoid unwanted delay of diagnosis and treatment.[12]

Our results indicate that long-distance journeys represent an important risk factor for VTE, requiring hospitalisation. The frequency of documented travel of 18.95% associated with DVT, makes travel an important risk factor for DVT.

In contrast, other studies reported air travel in less than 6% of VTE cases admitted to hospital.[5,10,11,13] In a study from Hawaii, recent air travel has been documented in 17–25% of patients admitted with VTE.[14–16]

In our group DVT occurred more frequently after long – distance travel by bus. This can be explained by a longer duration of bus travel (>24 h) and limited mobility, compared to travelling by plane or car. No DVT was recorded after travelling by train.

Many questions are still debatable and will require extended research. A recently published study, which was carried out on 1906 patients with travel-related thrombosis, concluded that the risk of venous thrombosis is 2-fold higher for all travellers and to the same extent for all modes of travel. In individuals with risk factors, the incidence of DVT is considerably higher, justifying the use of prophylactic therapy in these patients.

5. Conclusion

Long-distance travel represents a risk factor for DVT, but travel-associated thrombosis occurs mainly in travellers with risk factors for thrombosis.

The risk of DVT is increased by all modes of travel, not only by plane. In Romania, travel by bus makes about 50% of causes of travellers' thrombosis. This can be explained by the limited mobility and long duration of bus travel.

In sub-groups with additional risk factors, the DVT risk is highly increased and preventive measures are required. All travellers should be instructed regarding clinical signs of DVT, in order to avoid a long delay between onset of clinical signs and commencement of proper therapy.

References

1. J. Homans, *N. Engl. J. Med.* **250**, 148 (1954).
2. I.S. Symington and B.H.R. Stack, *Brit. J. Diseas. Chest.* **71**, 138 (1977).
3. J.M. Cruickshank, R. Gorlin and S. Jennett, *Lancet.* **2**, 497 (1988).
4. S. Mendis, D. Yach and A. Alwan, *Bull. WHO* **80**, 403 (2002).
5. M.M. Samama, *Arch. Intern. Med.* **160**, 3415 (2000).
6. P.L.F. Giangrande, *Brit. J. Haematol.* **117**, 509 (2002).
7. I. Martinelli, E. Taioli, T. Battaglioli, G.M. Podda and S.M. Passamonti, *Arch. Intern. Med.* **163**, 2771 (2003).
8. F. Paganin, A. Bourde, J.L. Yvin, R. Genin, J.L. Guijarro and A. Bourdin, C. Lassalle, *Aviat. Space. Environ. Med.* **74**, 1277 (2003).
9. T. Schwarz, G. Siegert, W. Oettler, K. Halbritter, J. Beyer, R. Frommhold, S. Gehrisch, F. Lenz, E. Kuhlisch, H.E. Schroeder and S.M. Schellong, *Arch. Intern. Med.* **163**, 2759 (2003).
10. E. Ferarri, T. Chevallier and A. Chapelier, *Chest.* **115**, 440 (1999).

11. R. Kraaijenhagen, D. Haverkamp and M.M.W. Koopman, *Lancet.* **356**, 1492 (2000).
12. Fr. Schneider, R. Siska-Ioana and J. Avram, *Clinical Physiology of the Venous System*, (Kluwer Academic Publishers 2003).
13. M. Wolde, R.A. Kraaijenhagen and J. Schiereck, *Thromb. Haemost.* **89**, 499 (2003).
14. A. Mercer and J.D. Brown, *Aviat. Space Environ. Med.* **69**, 154 (1998).
15. B. Eklof, R.L. Kistner, E.M. Masuda, B.V. Sonntag and H.P. Wong, *Dermatol. Surg.* **22**, 637 (1996).
16. B. Arfvidsson, R.L. Kistner, D.T. Sato, *Hematol. Oncol. Clin. North Am.* **14**, 391 (2000).

CHAPTER 7

MANAGEMENT OF ACUTE UPPER LIMB ISCHEMIA (ACULI) — RETROSPECTIVE STUDY

MIHAELA-IOANA PASZTORI, JECU AVRAM AND SIMONA MANCIU

County Hospital, Timisoara, Romania

CORINA STANCIULESCU

General Practitioner, Timisoara, Romania

Our aim was to evaluate the etiology producing acute upper limb ischemia, the treatment applied in these cases and the evolution of cases. In our retrospective study, we followed 29 cases of acute upper limb ischemia treated in the First Surgery Clinic, County Hospital Timisoara between the years of 2000 and 2006. The etiology of these 29 cases (16 males and 13 females, with an average age of 56 years) was the same as reported in other studies and papers, but the ratio between thrombosis and embolism was different from that reported by others. In our study we reported the following: thrombosis (8 cases — 26%), post-fibrillation embolism (10 cases — 35%), thrombosis associated with embolism (3 cases — 10%), traumas (5 cases — 17%), arterial steel syndrome and iatrogenic pseudo aneurysm secondary to arterio-venous fistula for haemodialysis (2 cases — 6%), micro-embolisation associated with spasm (1 case — 3%). Surgery was applied in 26 cases (3 cases treated only with vasodilators and Heparin). Fogarty balloon-catheter thrombectomy/thrombembolectomy was performed in 19 cases; brachial artery reconstruction was necessary in three cases. Immediate good results were obtained in 25 cases, re-thrombosis imposed re-intervention in two cases from which one evolved in a favourable way, and the other was followed by major amputation due to re-thrombosis. One case imposed a serious re-intervention with forearm amputation due to complex traumatic lesions. The patient with vascular coma and hemi paresis died due to evolution of neurological diseases.

1. Introduction

1.1. *Definition*

Acute upper limb ischemia is a disease with a relatively low incidence in vascular pathology (16–20%) characterised, like all the acute ischemia, by a sudden decrease in arterial blood flow from an extremity. The outcome of this phenomenon is a brand new symptomatology, or in case of a previous occlusive disease at that level, an aggravation of previous symptomatology.[3] The effects of sudden arterial occlusion depend on the state of collateral supply (collateral supply in a limb is usually inadequate unless a pre-existing occlusive disease determines the development of those collaterals). For a better recognition of an acute case, the international terminology recommends the following classification: acute ischemia-symptoms manifesting for less than 14 days, acute or chronic-worsening symptoms and signs with a duration under 14 days and chronic ischemia stable symptoms for more than 14 days.[1]

1.2. *Etiology*

The majority of acute limb ischemia is caused by acute thrombosis in a vessel with pre-existing atherosclerosis (60%) the embolism being recognised in just 30% of acutely ischemic limbs.[1] In other studies, the proportion between embolism and thrombosis as principal causes of acute limb ischemia has a different range (thrombosis 40–80%, embolism 30–60%). There are also other, less frequent causes including aortic dissection, trauma, iatrogenic injury (arterial cannulation), peripheral aneurysm, and injury caused by extreme cold, thoracic outlet syndrome, Raynaud's Syndrome.

1.3. *Clinical Diagnosis*

In acute ischemia, the severity of signs and symptoms is not proportional to the degree of ischemia. The initial sign is the pallor of the affected extremity due to the spasm of the distal arterial tree. Over the next hours, the extremity becomes light purple because of some degree of peripheral vasodilatation. As ischemia progresses the colour turns darker and becomes fixed. Clinical diagnosis depends on the classic six "P"s described first by G. H. Pratt later associating other "P"s:[4]

1. Pain (could be absent in complete ischemia)
2. Pallor (present also in most chronic cases)
3. Pulselessness (present also in most chronic cases)
4. Paraesthesia leading to anaesthesia (unable to feel touch on hand)
5. Paralysis (sign of a great importance indicating the need for urgent surgical treatment)
6. Poikilothermia/Perishing with coldness of the affected limb
7. Prostration
8. Poor capillary return
9. Buerger's test become positive

1.4. *Managing AcULI*

The assessment of these patients will be multidisciplinary because of the probable pre-existing coronary, cerebro-vascular and renovascular abnormalities. After a clear history and a rapid but complete examination, the next step is to identify the main underlying cause (thrombosis or embolism). The management and outcome of these two categories is different and thus, distinguishing between them is essential. The next step requires an evaluation of the severity of ischemia (acute-acute on chronic, incomplete-complete, and irreversible). The severity of ischemia is the key-factor when establishing the type of treatment that should be applied. There are two very distinct situations:

1. In case of an irreversible with a non-salvageable limb (more rarely encountered in these cases than in lower limb ischemia), the amputation is inevitable.

2. In cases with incomplete occlusion (viable limb) or a complete occlusion (acutely threatened limb), the management involves: initial intravenous analgesia and heparinisation (bolus of 5000 units of heparin followed by an infusion of 1000 units per hour) to prevent propagation of thrombus, treatment for associated cardiac disease and prompt angiography to plan further intervention.
3. The acutely threatened group needs expert vascular input and the treatment depends on the individual circumstances (thrombolysis, angiography with angioplasty, embolectomy or urgent arterial bypass). In these cases the followings options should be considered:

   a) In embolic disease — embolectomy directly (rarely used)/ indirectly (with Fogarty balloon) or intra-arterial thrombolysis using Streptokinase 5000u/hr + heparin 250u/hr or alternative thrombolytic agents-urokinase or tissue plasminogen activator/tPA),
   b) In thrombotic disease — intra-arterial thrombolysis with or without aspiration thrombectomy, angioplasty (PTA) used in stent restenosis or in chronic arterial stenosis or bypass surgery.[2]

2. Material and Methods

We studied 29 cases of acute upper limb ischemia treated in the First Surgery Clinic, County Hospital Timisoara, Romania in the period 2000–2006. Our aim was to emphasise the necessity and usefulness of surgical treatment in patients with upper limb ischemia even when the time between the onset of the symptoms and the admission in hospital is much longer than the "6–12 gold hours of lower limb acute ischemia".

3. Results

We registered an increasing number of patients in the years of 2002 and 2003, see Figure 1. As expected, the gender ratio shows a slight predominance of men, see Figure 2. The age and gender distribution

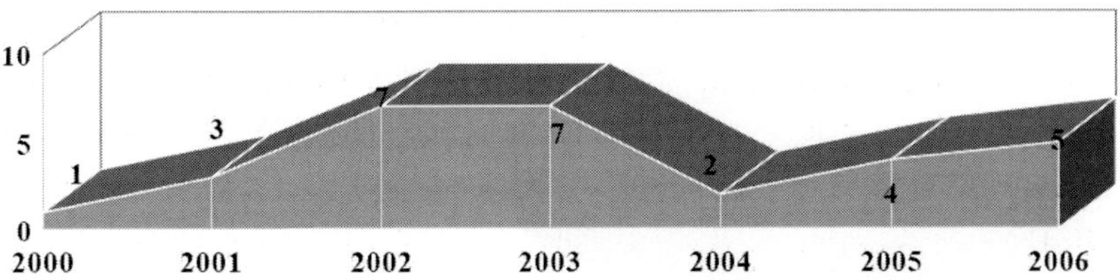

Figure 1. Number of patient's distribution per years of admittance.

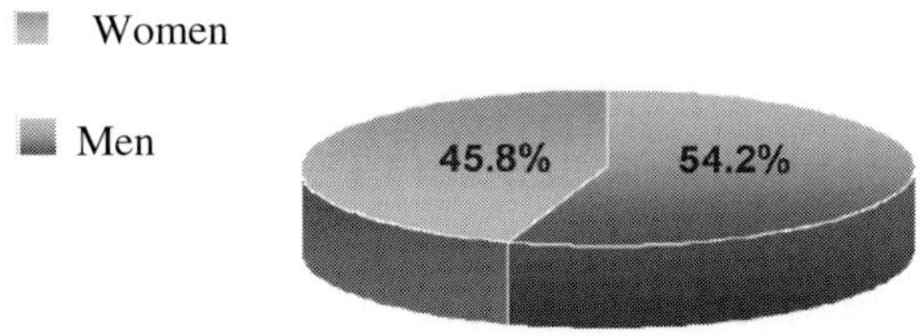

Figure 2. Gender ratio.

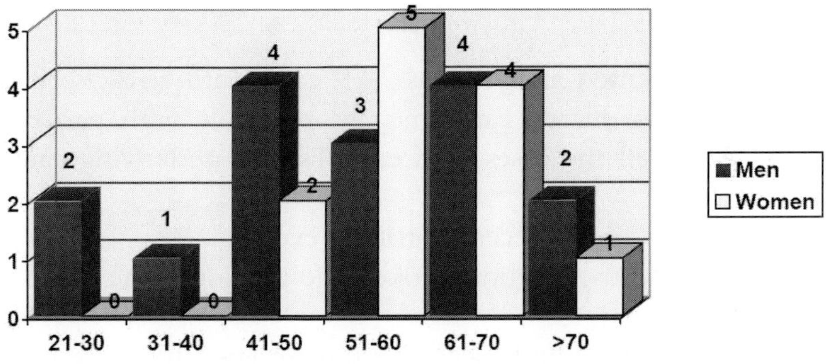

Figure 3. Age and gender distribution.

shows that in a single age interval (51–60 year) women are prevailing, see Figure 3.

The etiology of these 29 cases was the same as reported in other studies and papers, but the ratio between thrombosis and embolism was different from that reported by others. In our study the percentages were calculated for thrombosis (8 cases), post-fibrillation

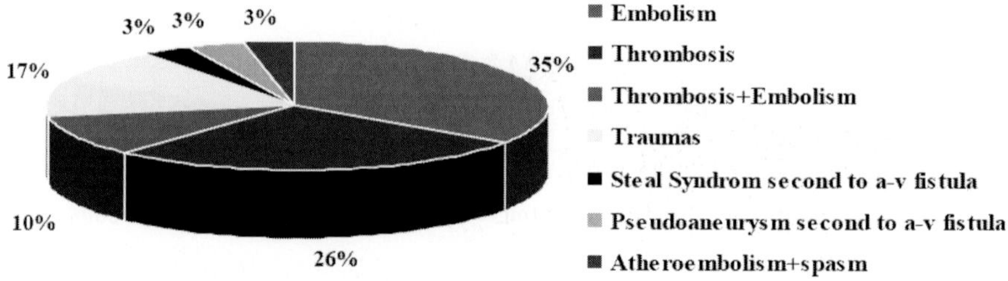

Figure 4. Distribution of patients according to etiology.

embolism (10 cases), thrombosis associated with embolism (3 cases with 1 case presenting an aneurysm of humeral artery in a chronic crouch holder), traumas (5 cases), arterial steel syndrome and iatrogenic pseudo aneurysm secondary to arterio-venous fistula for haemodialysis (2 cases) and micro-embolisation associate with spasm (1 case).

Associated pathologies registered in our retrospective study were represented by:

1. 10 cases presented arrhythmias — 9 cases with atrial fibrillation and 1 case of dilated cardiomegaly, probably with paroxysmal arrhythmias (all the cases with embolism as underlying cause of ischemia),
2. 2 cases with acute or recent coronary events,
3. 2 cases with ictus/cerebral stroke (one previous, one acute with left hemi paresis),
4. 4 cases with lower limb occlusive disease (two with pre-existing amputation),
5. 2 cases with chronic renal failure attending haemodialysis.

From all 5 cases with traumas, 3 were associated with complex lesions (2 with humerus fracture associated with fore-arm fracture, one of them associated also with venous and nervous lesions, the third associated with nervous concussion and muscular rupture).

The interval between the onset of symptomatology and admission in hospital was variable, majority of patients were treated after more

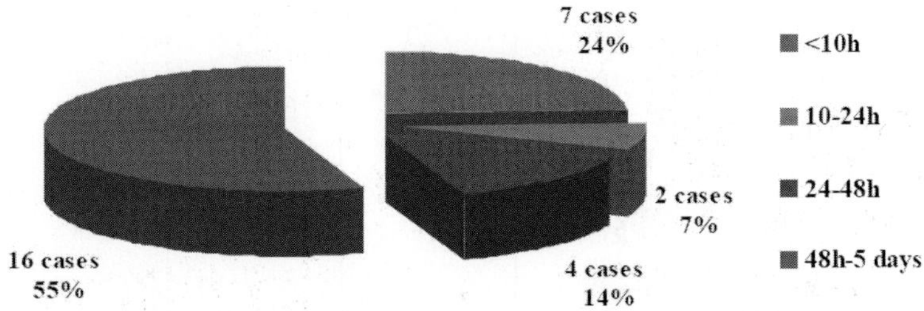

Figure 5. Distribution of patients according to the interval between onsets of symptomatology and admission into hospital.

than 48 hours from the onset, see Figure 5. Nevertheless, the percentage of irreversible lesions necessitating amputation was low as shown in the next diagrams.

Regarding the treatment, surgery was applied in 26 cases (3 cases treated only with vasodilators and Heparin). Fogarty balloon-catheter thrombectomy/thrombembolectomy was performed in 19 cases, brachial artery reconstruction being necessary in 3 cases. Other types of surgery applied were radial artery ligation in 2 cases, one pseudo aneurisms resection maintaining the arterio-venous fistula and in one case the interruption of arterio-venous fistula was necessary. All cases undergoing surgery were followed by Heparin administration, (see Figure 6).

Concerning the evolution, we registered immediate good results in 25 cases, re-thrombosis imposed re-intervention in 2 cases from which one evolved favourable, and the other was followed by major amputation due to re-thrombosis. One case imposed serious re-intervention with forearm amputation due to complex traumatic lesions. The patient with vascular coma and hemi paresis died due to the evolution of neurological diseases, see Figure 7.

4. Discussion

Regarding associated pathologies registered in our retrospective study, we noticed that more than one half of cases showed multiple

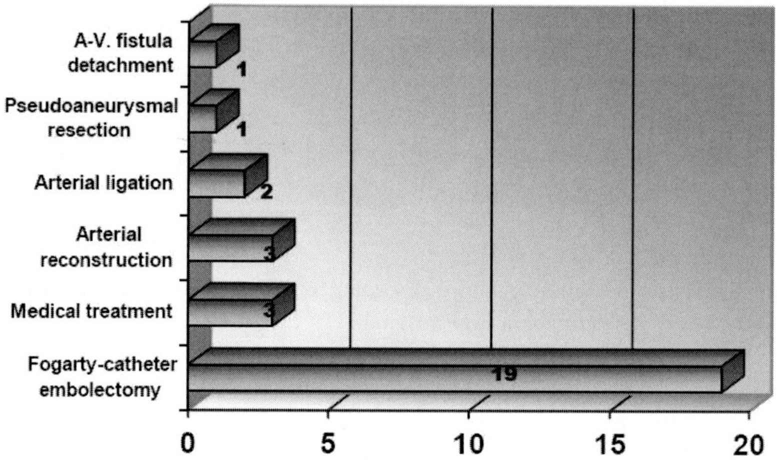

Figure 6. Repartition regarding the type of therapy.

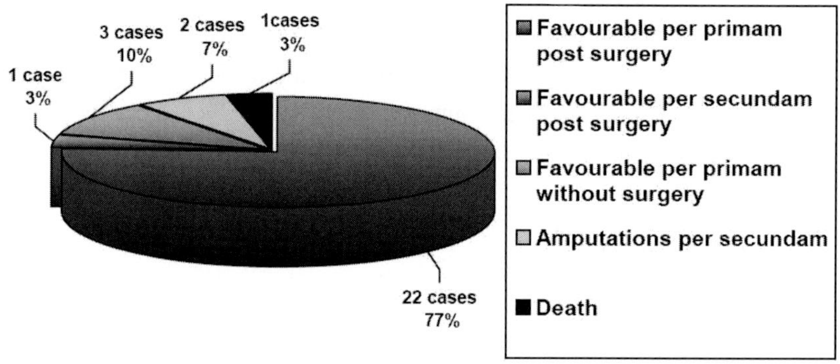

Figure 7. Distribution of patients' function of the evolution.

affections secondary to atherosclerosis and the process of aging (10 case — arrhythmias, 2 cases — acute or recent coronary events, 2 cases — ictus/cerebral stroke and 4 cases — lower limb occlusive disease). There are many theories related to atherogenesis, but most of them involve the alimentation habits. It is well-known that in the Banat region the alimentation habits consist of food rich in lipids (especially of animal origin). Changing the eating habits may reduce the number of patients with this pathology.

5. Conclusion

The disease is uncommon when compared to all the cases with vascular pathology under treatment in our clinic; the etiology was different, the main causes were embolism and thrombosis in elderly patients (over 45 years) with chronic pathologies (stroke, CHD, arrhythmias); arterial dissection, traumas, arterial compression were found in young active patients.

Most of the cases were treated by embolectomy with Fogarty catheter; the evolution depended on the associated pathology (1 death caused by stroke); local evolution was favourable even when the period between the installation of ischemia and the treatment was over 48 hours (until 72 hours), nevertheless the best results were obtained when addressability was good (under 48 hours).

In our country, surgical treatment is less expensive than both fibrinolyse and endovascular treatment.

References

1. M.J. Brescia, J.E. Cimino, K. Appel and B.J. Hurwich, *N. Engl. J. Med.* **275**, 1089 (1966).
2. M.T. Caps, *Semin. Vasc. Surg.* **11**, 227 (1998).
3. J.P. Elliot, J.H. Hageman and D.E. Szilagyi, *Surgery* **88**, 833 (1980).
4. H. Haimovich, *Vascular Surgery Principles and Techniques*, (Blackwell Science, Inc. 1996).

CHAPTER 8

INTERLOCK PRECISION MILLINGS IN COMBINED DENTURES TECHNOLOGY

LILIANA SANDU, CRISTINA BORTUN, FLORIN TOPOLĂ AND SORIN POROJAN

Faculty of Dental Medicine, "Victor Babeş" University of Medicine and Pharmacy Timişoara, Romania

Interlocks, approximally positioned drillings between the connected crowns, are essential to achieve reliable and durable combined dentures. The aim of the study was to detail laboratory procedures in achieving combined dentures with interlock milling systems. The paper also describes laboratory procedures of fixed and removable structures achieved by means of interlock milling systems in combination with stress distribution arms and ready-made attachments. The free precision milling techniques in wax and metallic frameworks are described step by step during the denture manufacturing procedures. The association of milling systems to other attachments guides the framework and acts against shear forces. Manufacturing of combined dentures with milling systems implies specific knowledge, precision in covering all the laboratory steps and an adequate endowment.

1. Introduction

The focus of partially edentulous patient's treatment is on restoration in accordance with the findings and long-term stabilisation of orofacial structures and functions. Often negative attitude towards removable partial dentures in comparison to other prosthodontic restorations is not justified. In cases of correct diagnosis, planning and design, removable partial dentures are an acceptable, functional replacement.[1–4]

There are differences between a clasp-retained partial denture and technically sophisticated combination denture in terms of function, aesthetics, as well as prosthodontic comfort. The last includes attachment dentures with milled support for mounting stress distribution arms and interlocks. The retention function is assumed by the attachment and the stress distribution arms and interlocks guide, support and splint of the denture.[5–8]

Most frequently used extra coronal attachments have the disadvantage that leverage may occur due to the decentralised positioning under masticatory force. Therefore, milling systems are necessary. The purpose of the study was to detail the laboratory procedures in achieving combined dentures with interlock milling systems.

2. Materials and Method

The paper describes laboratory procedures of fixed and removable structures achieving by means of interlock milling systems in combination with stress distribution arms and ready-made attachments. Therefore, a maxillary Kennedy class II edentation with one modification space was chosen.

The ready-made attachments, in the case of correct indication and proper processing, have a high degree of acceptance among patients. We used ancora extracoronal attachments (Bego, Bremen, Germany). These include an acrylic male part, which has to be attached in the primary wax up. The removable part of the denture is fixed on the primary frame by means of a replaceable female part, which is made of a resistant, permanently elastic acrylic and is anchored in the partial denture frame.

The free precision milling techniques in wax and metallic frameworks were made step by step during the denture manufacturing procedures. The split-cast working cast provided with removable dies was made of super-hard stone (class IV), see Figure 1. Teeth were made on this fully anatomic wax-up of the abutment, see Figure 2.

The path of insertion and removal was specified during the model analysis because it guides the milling procedures. Twisted tungsten

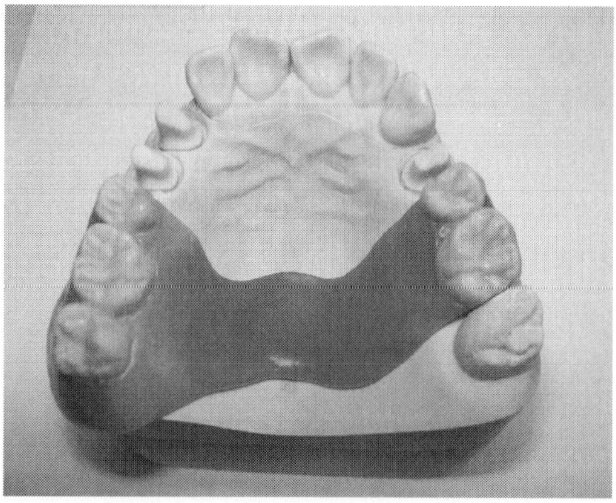

Figure 1. Split-cast working model with set-up.

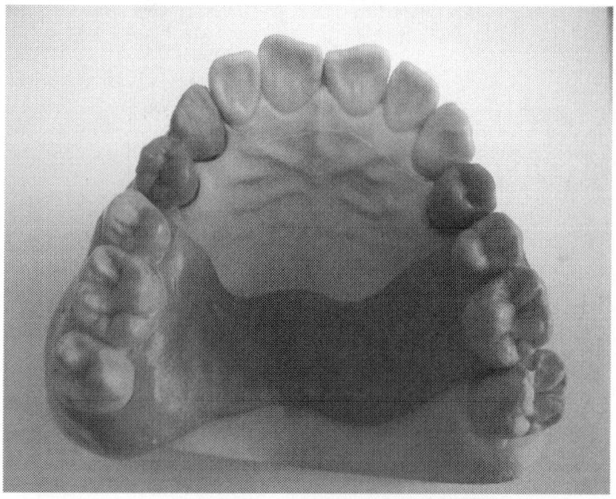

Figure 2. Full anatomic wax-up of the abutment teeth and set-up.

carbide wax cutters were used for efficient wax milling with Paraskop M (Bego, Bremen, Germany). It is helpful to pre-mark the position of ledges and drilled holes on the fully contoured wax-up. The height of the milled surface should be between 2.5 and 3 mm. The milling cutter was guided using very little pressure for counter-clockwise rotation milling at 1000–4000 rpm.

The milled cervical ledge ends in a chamfered groove shape, see Figure 3. This ledge shaping offers additional hygienic advantages.

Ledges milled at a right angle are more difficult to clean and also to cast. The ledges should be positioned approx. 1 mm above the marginal gingiva.

The occlusal shoulder was made at a right angle, see Figure 4. Shoulders positioned at a right angle to the path of insertion are best suited for the force transmission.

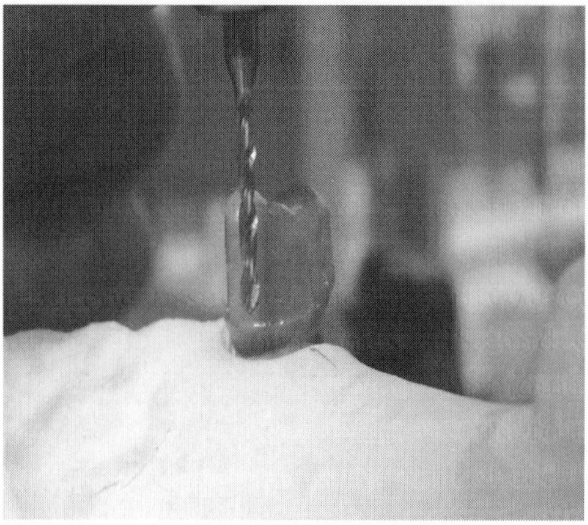

Figure 3. Wax milling of the cervical ledge.

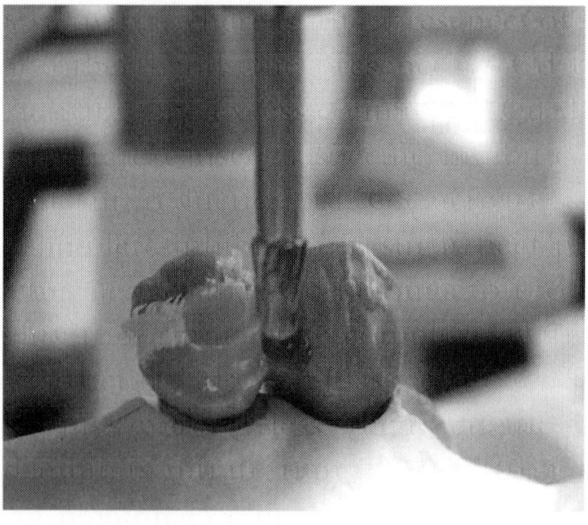

Figure 4. Wax milling of the occlusal shoulder.

It is extremely important that to position all milled surfaces and grooves at the same level and have approximately equal length, as far as possible. Friction grooves with a diameter of 1–1.2 mm were drilled as a 3/4 channel with a spiral drill, see Figure 5.

The wax patterns were sprued, see Figure 6, invested, see Figure 7, with Bellavest SH (Bego, Bremen, Germany) and cast, see Figure 8,

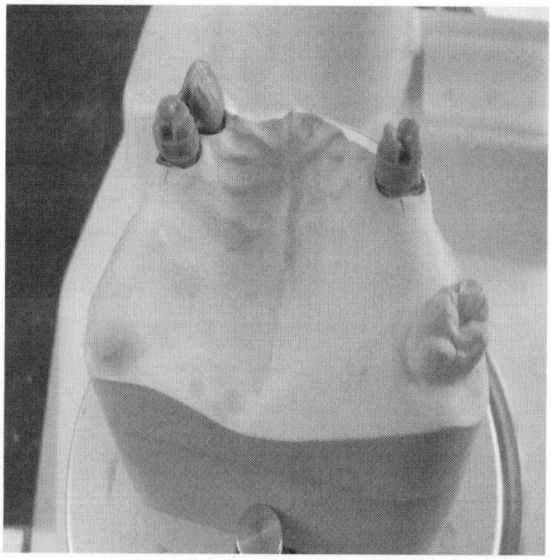

Figure 5. Completed wax milling with interlock and terminal grooves.

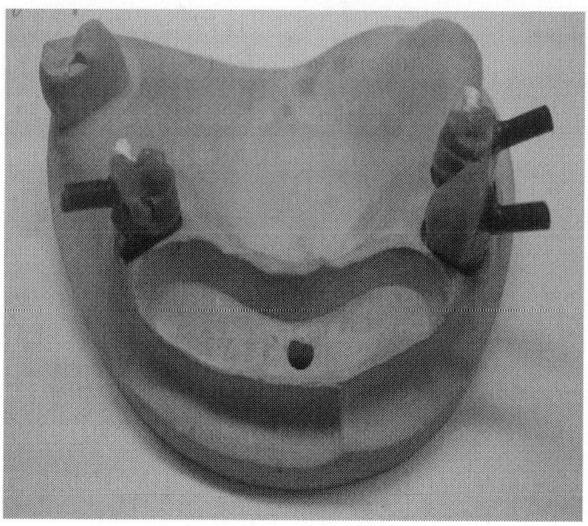

Figure 6. Sprued wax pattern.

Figure 7. Primary framework investing.

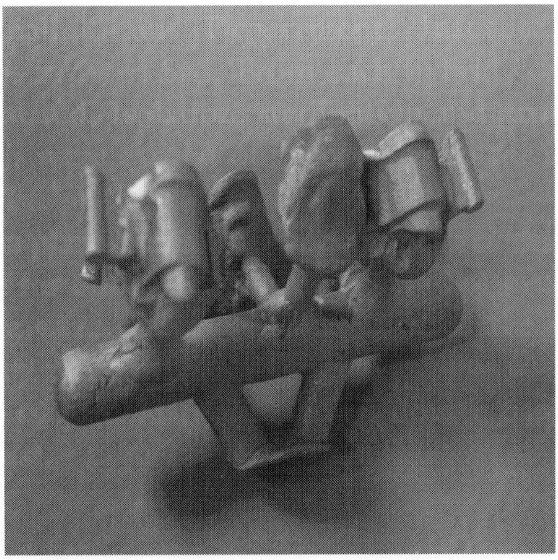

Figure 8. Primary framework after casting.

with Fornax T (Bego, Bremen, Germany) using a Ni-Cr alloy Wiron 99 (Bego, Bremen, Germany).

After deflasking, the primary frames were finished by sandblasting and grinding. The milling steps were made also in metal, using adequate milling cutters, at a rotational speed of 8000 rpm for clockwise rotation milling. The finish for the milled surfaces was provided with cotton wool, see Figure 9.

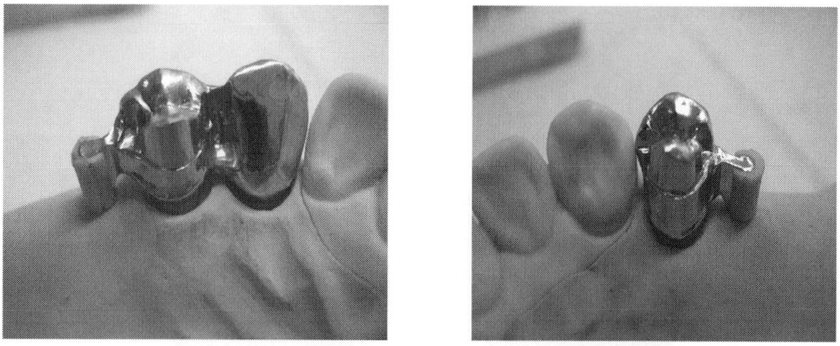

Figure 9. Finished primary frameworks on the working cast.

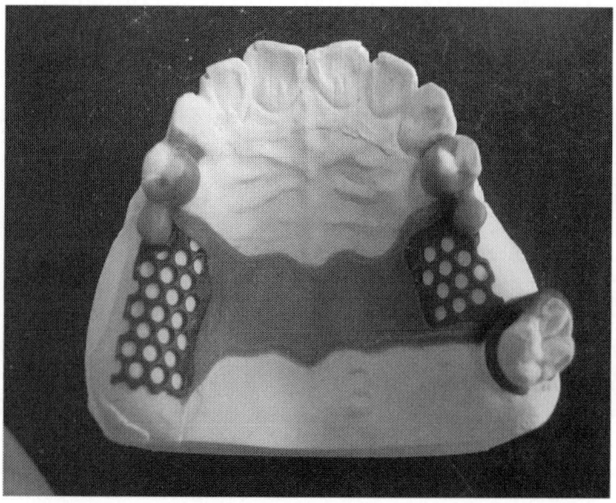

Figure 10. Maxillary framework wax-up.

The working cast were prepared for duplication. Silicone duplication is necessary for combined work. We used Wirosil duplicating flask system (Bego, Bremen, Germany). The refractory cast was poured using Wirofine investment (Bego, Bremen, Germany).

The partial denture framework was waxed on the duplicating cast, see Figure 10. The wax patterns were sprued, see Figure 11, invested with Wirofine (Bego, Bremen, Germany) and cast with Nautilus CC (Bego, Bremen, Germany) using a Co-Cr alloy Wironit extra hard (Bego, Bremen, Germany) after usual laboratory procedures.

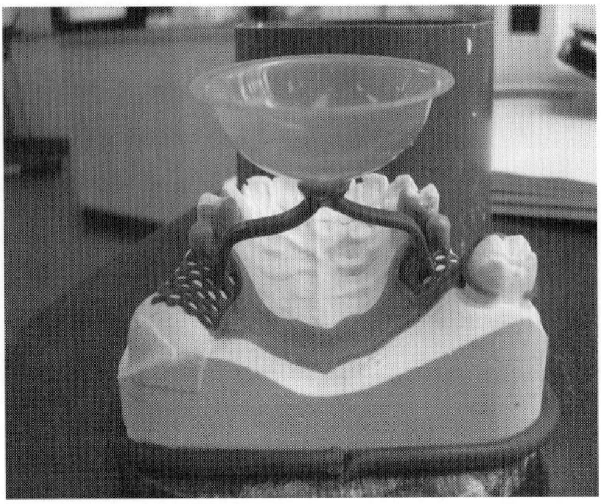

Figure 11. Sprued wax-pattern of the secondary framework.

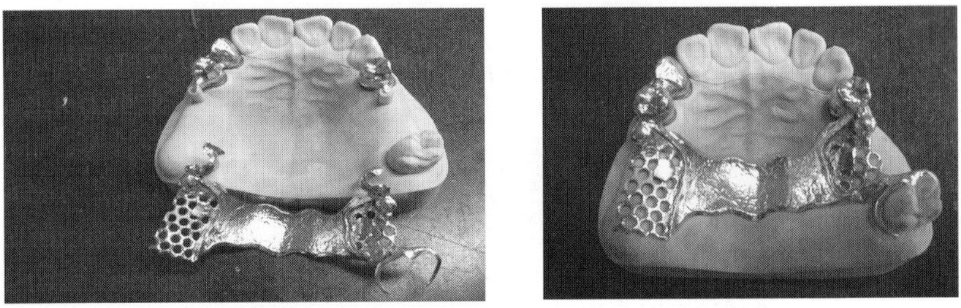

Figure 12. Finished frameworks of the combined denture placed on the working cast.

After deflasking, the frame was finished by sandblasting, grinding and electrolytic polishing. The fitting accuracy was checked on the working cast and the frame was rubber-polished.

3. Results and Discussions

Finally, the fit between the primary and secondary frame was checked and the acrylic female parts were inserted in the secondary frame, see Figure 12.

Milled ledges and interlocks on primary parts serve as retaining and supporting elements in combination with ready-made attachments.

They act as guide surfaces for insertion and removal and prevent saddle from lifting off. The interlock also prevents stress distribution arm from bending up, ensures secure seating of the denture and acts against shear forces.

4. Conclusion

Manufacturing of combined dentures with milling systems implies specific knowledge, precision in covering of all the laboratory steps and an adequate endowment. Modern equipment and materials make construction easier, but they do not solve the problems related to planning and design.

References

1. H.H. Caesar and K.M. Lehmann, *Removable Partial Denture — Fundaments, Design and Achievement*, (Verlag Neuer Merkur GmbH, Munchen, 2002).
2. C. Borţun and L. Sandu, Design-ul protezelor parţiale mobilizabile, Lito UMF Victor Babes (2005).
3. M. Walter, Planing and treatment strategy at fixed/removable partial dentures I — direct retainers, Universitätsklinikum Carl Gustav Carus Dresden (2002).
4. H. Wulfes, Precision milling and partial denture construction, academia dental, *International School*, Bego, Germany (2004).
5. A.B. Carr, G.P. Mc Givney and D.T. Brown, Mc Cracken's removable partial prosthodontic, Mosby (2005).
6. R.D. Phoenix, D.R. Cagna and C.F. De Freest, Stewart's clinical removable partial prosthodontics (2003).
7. C. Borţun, M. Leretter and L. Sandu, Tehnologia protezelor parţiale mobilizabile, Vol. I and II, *Eurobit*, Timişoara (2002).
8. M. Saito, K. Notani, Y. Miura and T. Kawasaki, *J. Oral. Rehabil.* **30**, 482 (2003).

4. Conclusion

PART 2

THE AGRICULTURAL AND FOOD TECHNOLOGY ISSUES

CHAPTER 9

ANTIOXIDANT ACTIVITY OF WILD AND CULTIVATED *ALLIUM* SPECIES FROM VOJVODINA

DUBRAVKA ŠTAJNER AND BORIS POPOVIĆ

Faculty of Agriculture, University of Novi Sad, Novi Sad, Serbia

Cultivated and wild *Allium* species were investigated to evaluate their antioxidant properties. The antioxidative enzymes and scavenger activities, quantities of non-enzymic antioxidants, malonyldialdehyde and HO$^{\bullet}$ radical were determined. The obtained results suggest that bulbs and leaves of cultivated *A. vineale L.*, *A. fistulosum L.* and *A. nutans L.* could be the promising sources of raw material for producing non-toxic natural antioxidants for food, pharmaceutical and cosmetic industries. However, further investigation is needed in that area. Cultivation of some wild sorts, such as *A. flavum L.*, *A. schoenoprasum L.* and *A. ursinum L.*, may be the focus of future research efforts in order to produce Allium sorts with strong antioxidant quality.

1. Introduction

Throughout the recorded history, *Alliums*, especially garlic and onion, played rich diverse commercial, culinary and mystic roles. The ancient Egyptian and Middle Eastern cultures used garlic as a currency. Around 2500 BC, 15 pounds of garlic could purchase a healthy slave. The culinary delights of garlic over manna were extolled by followers of Moses in the Old Testament. In the present time, the interest in research of other *Allium* species is also observed.[1] Today, garlic and onion are used for their flavour, aroma and taste properties, being prepared domestically or forming basic materials for a variety of food manufacturing processes. Onions were among the earliest vegetables to be processed, canned, dried and frozen.[2]

Many epidemiological studies have suggested that certain natural foods could prevent the development of different diseases. Garlic and onion are such natural foods. They have a variety of pharmacological effects including tumour cell growth inhibition and chemo-preventive activity.[3] Much of the data about human use came from the reports of lowered rates and risks of disease (such as cancer) in people with relatively high levels of garlic or other *Alliums* consumption. People also use garlic and onion to help with several different types of ailments, such as high cholesterol, high blood pressure, excess blood clotting and coagulation, atherosclerosis, inflammation, bacterial and fungal infections. Garlic is also a functional food product composed of numerous macronutrients, vitamins,[2] organosulphur active compounds, proteins,[4] lipids, trace elements as Se,[5] flavorous,[6] enzymes[7] and different antioxidants.[8]

Our previous study on antioxidant ability of different *Allium* species[9–11] showed that they possess antioxidant abilities in all organs, but especially in leaves. Therefore, the aim of this study was to explore leaves antioxidant activities of some wild and cultivated *Alliums* in order to point to species which could be the sources *of non-toxic* natural antioxidants for food, pharmaceutical and cosmetic industries.

2. Experimental Methods

2.1. *Plant Material*

Allium plants were collected in the flowering phase. For the experiment, fresh leaves of both cultivated (*Allium nutans* L., *Allium fistulosum* L., *Allium vineale* L., *Allium psekemense* B. Fedtsch, *Allium cepa* L., *Allium sativum* L.) and wild (*Allium flavum* L., *Allium sphaerocephalum* L., *Allium atroviolaceum* Boiss, *Allium schenoprasum* L., *Allium vineale* L., *Allium ursinum* L., *Allium scorodoprasum* L., *Allium roseum* L. and *Allium subhirsutum* L.) species were used. Voucher herbarium specimens are deposited at the Institute of Biology, University of Novi Sad.

2.2. *Methods*

One gram of fresh plant material was grounded with quartz sand in a cold mortar. The ground material was suspended in 5 ml 1 mol/L K_2HPO_4 at pH 7.0. After 10 minutes of centrifugation at 4°C and 15 000 g, the aliquots of the supernatant were used for super-oxide dismutase (SOD) activity measurements. All measurements were made in triplicate. The values are expressed as mean ± standard error.

Prior to measurement of the enzyme activity we have added 20 μL of Tsuchiachi solution (chloroform/ethanol 3/5) to the supernatant. The SOD activity was determined in aliquots by the method of Misra and Fridovics[12] based on the inhibition of transformation of adrenaline to adrenochrome at pH 10.2.[13]

Carotenoids were extracted with acetone and determined spectrophotometrically using molar extinction coefficients according to Wettstein.[14] The amount of reduced glutathione (GSH) was determined with Ellman reagent at 412 nm.[15]

Lipid peroxidation (LP) was determined by the thiobarbituric acid (TBA) method; values were given as equivalent amounts of malonyldialdehyde (MDA); the calibration curve was prepared with malonyldialdehyde bis-diacetal.[16] Hydroxyl radical was determined by the inhibition of deoxyribose degradation.[17]

The influence of the phosphate buffer (pH 7) *Allium* extract on hydroxyl radical (HO) formation was studied by electron spin resonance (ESR) using a spin trapping method.[18] The scavenging activity of the extract was estimated by the percentage decrease of the relative intensity of signal of DMPO-OH radical adduct, with reference to the control without extract. All experiments were repeated three times.

3. Results and Discussion

The SOD activities and GSH quantities in leaves and bulbs and carotenoid content in leaves of different *Alliums* are presented in Table 1. All investigated leaves and bulbs exhibited SOD activity. In leaves, SOD activity ranged from 2.51 U/mg proteins in *A. ursinum* to 77.46 U/mg proteins in *A. schenoprasum* and in bulbs

Table 1. SOD activities and GSH quantities in leaves and bulbs and carotenoid content in leaves of different *Alliums*

Allium sort	SOD U/mg protein (leaves)	SOD U/mg protein (bulbs)	GSH µmol/mg protein (leaves)	GSH µmol/mg protein (bulbs)	Carotenoids mg/g (leaves)
A. sativum L.	52.47 ± 7.11	127.11 ± 12.62	0.215 ± 0.009	0.184 ± 0.002	2.57 ± 0.00
A. vineale L. (cultivated)	39.95 ± 2.34	20.82 ± 0.00	0.200 ± 0.003	0.186 ± 0.002	3.04 ± 0.01
A. vinaele L. (wild)	3.63 ± 0.52	3.41 ± 0.67	0.108 ± 0.001	0.142 ± 0.002	1.30 ± 0.03
A. sphearocephalum L.	8.91 ± 2.52	8.52 ± 1.97	0.453 ± 0.004	0.239 ± 0.002	1.91 ± 0.02
A. scorodoprasum L.	45.54 ± 23.88	39.89 ± 8.21	0.182 ± 0.016	0.940 ± 0.048	0.25 ± 0.01
A. atroviolaceum Boiss.	7.46 ± 1.51	2.24 ± 0.76	0.399 ± 0.011	0.139 ± 0.000	0.71 ± 0.01
A. cepa L.	2.69 ± 0.76	5.20 ± 0.62	0.127 ± 0.001	0.129 ± 0.003	1.92 ± 0.01
A. pskemenese L.	4.08 ± 0.84	16.18 ± 6.26	0.177 ± 0.004	0.336 ± 0.011	0.66 ± 0.01
A. fistulosum L.	62.20 ± 3.85	43.03 ± 4.93	0.497 ± 0.008	0.219 ± 0.006	2.87 ± 0.03
A. schenoprasum L.	77.46 ± 5.98	57.61 ± 4.79	0.669 ± 0.009	0.167 ± 0.001	2.14 ± 0.01
A. nutans L.	20.27 ± 3.27	3.73 ± 1.57	0.125 ± 0.004	0.072 ± 0.002	2.24 ± 0.01
A. ursinum L.	2.51 ± 1.21	6.42 ± 1.25	0.148 ± 0.003	0.096 ± 0.004	3.99 ± 0.01
A. flavum L.	10.62 ± 2.02	1.24 ± 0.9	0.146 ± 0.002	0.112 ± 0.002	1.10 ± 0.01
A. roseum L.	25.91 ± 3.36	43.36 ± 5.23	0.646 ± 0.08	0.279 ± 0.05	0.874 ± 0.02
A. subhirsutum L.	18.45 ± 2.18	14.75 ± 1.05	0.525 ± 0.13	0.646 ± 0.08	1.428 ± 0.03

from 1.24 U/mg proteins in *A. flavum* to 127.11 U/mg proteins in *A. sativum*. SOD present in leaves and bulbs remove O_2^- from the compartments where O_2^- radicals are formed including chloroplast and mitochondria, controlling oxidative stress in plants.[19]

The GSH quantity in leaves ranged from 0.108 nmol/mg protein (*A. vineale*, wild) to 0.669 nmol/mg (*A. schenoprasum*) and in bulbs from 0.072 nmol/mg in *A. nutans* to 0.940 nmol/mg in *A. scorodoprasum*. The quantity of GSH, a powerful non-enzymic antioxidant, was the highest in leaves and lowest in bulbs of the examined *Alliums*, thus indicating high antioxidant capacity of leaves. A high GSH quantity is also beneficial for lipid peroxide balance in tissue.[20]

Carotenoid content was detected in leaves of all investigated *Alliums*, but there was no evidence of its presence in bulbs. The highest carotenoid content (3.99 mg/g) was observed in leaves of *A. ursinum* and the lowest in *A. scorodoprasum*, 0.25 mg/g. It is well known that carotenoids among flavonoids and other polyphenol compounds scavenge lipid peroxyl radicals resulting of formation a relatively stable antioxidant radical, which react slowly with substrate (LOOH) and so reduced LP.[9]

The accumulation of OH radicals and MDA, the most toxic species and main products of lipid membrane peroxidation, is shown in Table 2. The lowest ·OH quantity was observed in leaves of *A. pskemenese* (0.07 nmol/mg) and bulbs of *A. cepa* (0.59 nmol/mg) and the highest in leaves of *A. subhirsutum* (39.27 nmol/mg) and bulbs of *A. roseum* (96.23 nmol/mg). The lowest MDA quantity was observed in leaves of *A. fistulosum* (4.98 nmol/mg) and bulbs of *A. cepa* (12.80 nmol/mg) and the highest in leaves of *A. scorodoprasum* (130.19 nmol/mg) and in bulbs of *A. scorodoprasum* (550.94 nmol/mg).

According to our results, the MDA accumulation could be correlated with the high quantity of toxic $HO^\bullet$ resulting from membranes deterioration. The leaves exhibited the lowest lipid peroxidation due the accumulation of relatively low $HO^\bullet$ quantities[21] and presence of carotenoids.

The ·OH radical scavenging capacity of *Allium* leaf-extracts is presented in Table 3. All investigated *Alliums* showed huge scavenging

Table 2. Quantities of ·OH and MDA in leaves and bulbs of different *Alliums*

***Allium* sort**	**·OH nmol/mg protein (leaves)**	**OH nmol/mg protein (bulbs)**	**MDA nmol/mg protein (leaves)**	**MDA nmol/mg protein (bulbs)**
A. sativum L.	2.10 ± 0.29	3.73 ± 0.05	24.55 ± 1.02	42.73 ± 0.92
A. vineale L. (cultivated)	2.32 ± 0.05	2.21 ± 0.37	13.19 ± 0.51	26.39 ± 0.27
A. vineale L. (wild)	2.56 ± 0.03	16.21 ± 0.39	30.67 ± 0.47	148.96 ± 0.49
A. sphearocephalum L.	1.29 ± 0.11	56.54 ± 1.98	50.83 ± 0.73	368.92 ± 7.61
A. scorodoprasum L.	25.52 ± 1.20	39.12 ± 1.86	130.19 ± 1.24	550.94 ± 12.72
A. atroviolaceum Boiss.	3.21 ± 0.20	13.15 ± 0.09	67.09 ± 1.02	179.45 ± 1.13
A. cepa L.	0.82 ± 0.16	0.59 ± 0.03	7.61 ± 0.36	12.80 ± 0.24
A. pskemenese L.	0.07 ± 0.02	17.52 ± 0.34	7.12 ± 0.77	13.29 ± 3.84
A. fistulosum L.	0.20 ± 0.02	2.86 ± 0.32	4.98 ± 0.34	26.38 ± 0.39
A. schenoprasum L.	4.06 ± 0.24	4.08 ± 0.48	36.85 ± 0.60	44.75 ± 1.46
A. nutans L.	0.30 ± 0.04	1.16 ± 0.07	12.40 ± 0.16	14.28 ± 0.23
A. ursinum L.	0.38 ± 0.03	6.56 ± 0.17	7.27 ± 0.21	77.90 ± 6.62
A. flavum L.	0.43 ± 0.13	9.53 ± 0.02	7.74 ± 0.40	89.51 ± 4.10
A. roseum L.	24.49 ± 10.52	96.23 ± 21.11	35.70 ± 7.41	92.82 ± 18.82
A. subhirsutum L.	39.27 ± 7.13	65.21 ± 8.03	37.12 ± 12.03	73.69 ± 19.52

Table 3. Scavenging activity in leaves of different *Alliums*

Allium sort	Scavenging activity (%)
A. sativum L. (cultivated)	90.36
A. flavum L. (wild)	94.30
A. psekemense B.Fedtsch	50.22
A. fistulosum (autumn)	82.15
A. nutans L. (cultivated)	70.97
A. fistulosum L.	87.09
A.. flavum L. (cultivated)	69.68
A. roseum L.	84.61
A. subhirsutum L.	81.92

ability. The ESR data demonstrate that phosphate buffer extracts possess similar $HO^{\bullet}$ scavenging activities that is crucial because hydroxyl radicals are the major active oxygen species causing lipid oxidation and enormous biological damage.[8]

The highest scavenger activity was observed in wild *A. flavum* extract (94.3%). Other results concerning *A. flavum* support this assessment because the quantities of $^{\cdot}OH$ and MDA were low (Table 2). The scavenging activity of *A. sativum* extract was also high (90.36%) and this finding is in agreement with the results of a number of authors[22–24] showing antioxidant and pharmacological activities of garlic and onions. Our results are yet another confirmation that antioxidant and scavenger activities influence the pharmacological activity of garlic.

4. Conclusion

Our results suggest that due to high antioxidant and scavenger capacities bulbs and leaves of cultivated *A. vineale*, *A. fistulosum* and *A. nutans* L. could be the promising sources of raw material for producing non-toxic natural antioxidants for food, pharmaceutical and cosmetic industries. However, further investigations are needed in that area. Cultivation of some wild sorts, such as *A. flavum* L., *A. schoenoprasum* L. and *A. ursinum* L., may be the focus of future research efforts aimed at producing *Allium* sorts with strong antioxidant quality.

Acknowledgements

This investigation was supported by the Ministry of Science, Technology and Development, Republic of Serbia. The authors thank Professor Ruzica Igic from the Department of Biology and Ecology, Faculty of Science in Novi Sad for providing plant material and Professor Jasna Čanadanović-Brunet from Faculty of Technology in Novi Sad for ESR data.

References

1. H.D. Reuter, *Phytomedicine* **2**, 73 (1995).
2. J.L. Brewster and H.D. Rabinowitch, *Onion and Allied Crops* **74** (CRC Press, 1990).
3. H.P. Koch and L. Lawson, *The Science and Therapeutic Application of Allium Sativum L. and Related Species* (Williams and Wilkins, 1996).
4. E. Block, J.Z. Gillies, C.W. Gillies, A.A. Bazzi, D. Putman, L.K. Revelle, A. Wall, D. Wang and X. Yhang, *J. Am. Chem. Soc.* **118**, 7492 (1996).
5. C. Ip and D.J. Lisk, *The Attributes of Selenium — Enriched Garlic in Cancer Prevention, Prevention and Treatment* (Plenum Press, 1996).
6. P.C. Hollman, J.M. van Trijp, M.N. Buystman and M.S. van der Gaag, *Febs. J.* **418**, 152 (1998).
7. J.E. Roberts and V.E. Tyler, *Tylers Herbs of Choice* **155** (The Haworth Herbal Press, 1999).
8. N. Ide and Y. Itakura, *Phytother. Res.* **10**, 340 (1996).
9. D. Stajner, N. Milic-DeMarino, J. Canadanovic-Brunet and R. Igic, *Pharm. Pharmacol. Lett.* **8**, 161 (1998).
10. D. Stajner, N. Milic and J. Canadanovic-Brunet, *Phytother. Res.* **13**, 333 (1999).
11. D. Stajner, N. Milic-DeMarino, J. Canadanovic-Brunet and M. Popovic, *Phytother. Res.* **16**, 484 (2002).
12. H.P. Misra and I.J. Fridovics, *J. Biol. Chem.* **247**, 3170 (1972).
13. B. Matkovics, R. Novak, D.U.C. Hanh Huang, I. Syabo, Sz.I. Varga and A. Yelesna, *Comp. Biochem. Physiol.* **56B**, 31 (1977).

14. D. Wettstein, *Exp. Cell Res.* **12**, 427 (1957).
15. I. Sedlak and R.H. Lindsay, *Anal. Biochem.* **16**, 205 (1968).
16. Z.A. Placer, N. Custman and B.C. Hohnson, *Anal. Biochem.* **16**, 359 (1968).
17. K.H. Cheesman, A. Beavis and H. Esterbaurer, *Biochem. J.* **232**, 649 (1988).
18. K. Hiramoto, N. Ojima, K.I. Sako and K. Kikugawa, *Biol. Pharm. Bull.* **19**, 558 (1996).
19. R.G. Alsher, N. Ertuk and L.S. Heath, *J. Exp. Bot.* **50**, 1331 (2002).
20. G. Filomeni, G. Rotilio and M.R. Ciriolo, *Biochem. Pharmacol.* **64**, 1059 (2002).
21. B. Halliwell and J.M.C. Guteridge, *Free Radicals in Biology and Medicine* **24** (Claredon Press, 1989).
22. A. Helen, K. Krishnakumar, P.L. Vijayammal and K.T. Augusti, *Toxicol. Lett.* **116**, 61 (2000).
23. W.H. Briggs, J.D. Folts, H.E. Osman and I.L. Goldman, *J. Nutr.* **131**, 2619 (2001).
24. J.D. O'Reilly, A.I. Mallet, G.T. McAnlis, I.S. Young, B. Halliwell, T.A. Sanders and H. Wiseman, *Am. J. Clin. Nutr.* **73**, 1040 (2001).

CHAPTER 10

ANTIOXIDANT SYSTEMS IN SOME RARE AND ENDANGERED PLANT SPECIES FROM THE PANNONIAN SALINE MEADOWS AND GRASSLANDS

ĐORĐE MALENČIĆ, DUBRAVKA ŠTAJNER, MILAN POPOVIĆ, ALEKSA KNEŽEVIĆ, LJILJANA NIKOLIĆ AND DEJANA DŽIGURSKI

Faculty of Agriculture, University of Novi Sad, Novi Sad, Serbia

In order to establish the level of oxidative stress, antioxidant status was evaluated in several rare and endangered plant species from southern parts of the Pannonian plain, in Vojvodina province. The quantities of reactive oxygen species and malonyldialdehyde, activities of several antioxidant enzymes and content of reduced glutathione were determined. Also, the DPPH-radical scavenging activity of plant extracts was measured. Several different systems were found to be involved in plant defence against oxidative stress. It seems that antioxidant status could serve as a suitable biochemical criterion for evaluation of the level of stress in plants.

1. Introduction

In northern parts of Serbia (Vojvodina province) a large number of plant taxa from endemic and subendemic groups of the Pannonian plain are still present today. Gradual deterioration of natural habitats due to cultivation of land, forest management, plant collecting, ploughing, irrigation etc., resulted in decrease in their number.[1,2] Among intensive anthropogenic influences, land-reclamation measures are predominant in order to intensify agricultural production on saline (solonchak) meadows, swamps and marshes, and to afforest steppe habitats. Besides that, a part of the problem is also the uncontrolled collecting of rare plants, burning of steppe habitats, forming of garbage deposits and plant diseases.[3] In the Pannonian parts of Serbia a gradual deterioration of natural habitats resulted in a decrease in the number of endemic plant species. In this study, we investigated the Pannonian endems *Lepidium cartilagineum* (J. May) Thell., *Plantago schwarzenbergiana* Schur, *Rorippa kerneri* Menyh., *Rhinanthus borbasii ssp. borbasii* (Dörfl.) Soó, *Lotus siliquosus* L. and *Trigonella procumbens* (Besser) Rchb. growing on saline meadows and Pannonian grasslands. They can be found in Vojvodina province, in Hungary, Romania and east- and south Ukraine. Some of them, such as *P. schwarzenbergiana*, belong to the group of plant species of international importance for preservation of biodiversity while others show some medicinal properties — antiseptic (*L. cartilagineum*), as expectorants, for skin wounds treatment and in digestive problems (*P. schwarzenbergiana*) etc.

Many environmental factors provoke stress in plants which may lead to reduction and activation of O_2 on a cellular level. Factors such as UV light and other forms of radiation, photo oxidation, air pollution, drought, herbicides, heavy metals, pathogen invasion, certain injuries, hyperoxia, ozone, temperature fluctuations and other stresses are known to induce free radical formation in most aerobic organisms.[4,5] Electron acceptors such as molecular oxygen react easily with free radicals to become radicals themselves [reactive oxygen species (ROS)]. The primary derivatives of oxygen, such as superoxide anion radicals (O_2^-), hydrogen peroxide (H_2O_2), hydroxyl radicals (OH) and the singlet oxygen ($^1O_2$), play an important role in mediating ROS-related effects. Reactive species with short life cycle generated *in situ* can react with non-radicals and produce chain reactions. The ROS have been implicated in a number of physiological disorders in plants.[6] ROS inhibit chloroplast development, decrease seed viability and root growth and stimulate leaf abscission and desiccation. The main cellular components susceptible to damage by free radicals are polyunsaturated fatty acids in membranes, proteins, carbohydrates, nucleic acids and pigments such as chloroplasts or carotenoids.[7] Thus, the antioxidant defence systems have co-evolved with aerobic metabolism to counteract oxidative damage from ROS. Antioxidant enzymes such as superoxide-dismutase (SOD), plant peroxidases (Px) and catalase (Cat), as well as some other natural antioxidants such as reduced glutathione (GSH), phenolics (tocopherols, flavonoids, phenolic acids), pigments etc., constitute a system that keeps ROS at low steady-state concentrations in cells and tissues and prevent oxidative situations. Recent investigations showed that the antioxidant properties of plants could be correlated with oxidative stress defence and different human diseases including cancer, atherosclerosis and the aging process.[8]

The aim of this study was to investigate oxidative stress parameters and antioxidant status of rare and endangered plant species from saline meadows in order to evaluate influence of the environment on their natural habitats.

2. Material and Methods

The above-ground parts of selected plant species were chosen for the experiment. The plant material was collected in their natural habitats (saline meadows) during full blossoming in summer 2005, at different localities in Vojvodina province in the southern part of the Pannonian plane.

Quantities of superoxide ($^{\bullet}O_2^-$) and hydroxyl ($^{\bullet}OH$) radicals and malonyl-dialdehyde (MDA) and reduced glutathione (GSH) contents, as well as the activities of two antioxidant enzymes: superoxide-dismutase (SOD; EC 1.15.1.1) and guaiacol-peroxidase (GPx; EC 1.11.1.7) were determined in the leaves of investigated plant. In addition, the potential antioxidant activity of the test sample has been assessed based on scavenging activity of the stable DPPH free radicals.[9]

Superoxide radical was studied by the inhibition of adrenaline auto oxidation.[10] Hydroxyl radical was measured by the inhibition of deoxyribose degradation.[11] The lipid peroxidation (LP) was measured as malonyl-dialdehyde (MDA) production at 532 nm with thiobarbituric acid (TBA).[12] The GSH quantity was determined with Elman reagent.[13]

Enzyme assay: one gram of plant material was homogenized with 5 mL 0.1 mol dm^{-3} K_2HPO_4 at pH 7.0. After centrifugation for 10 min at 4°C and 15 000 g, aliquots of the supernatant were used for enzyme activity measurements. SOD activity was determined by the method based on the autocatalytic transformation of epinephrine-adrenochrome at pH 10.2.[10] GPx activity was measured using the method of Simon *et al.*[14]

3. Results and Discussion

The results obtained from the study of antioxidant properties of selected rare and endangered plant species from the Pannonian plain are shown in Tables 1 and 2. Our results showed that the examined plant species were partially susceptible to the action of toxic oxygen species. The presence of ROS was established in all samples probably due to high outside temperatures and water deficit in the mid-July when the plant material was collected (Table 1). In addition to normal metabolic activity, ROS can result from cellular exposure to various environmental stimuli.[6] Unfavourable environmental factors, such as water deficit and heath, may cause the disturbance in the process of molecular oxygen reduction when toxic and highly reactive oxygen species are being produced. O_2^- and $^{\bullet}OH$, in above-ground parts of *Rhinantus borbasii ssp. borbasii* were produced

Table 1. O_2^-, OH and MDA quantity in above-ground parts of selected rare and endangered plant species from the Pannonian plain

Specimen	O_2^- quantity (μmol g^{-1} fr. w.)	OH quantity (nmol g^{-1} fr. w.)	MDA quantity (nmol g^{-1} fr. w.)
Lepidium cartilagineum	4.47	629.65	43.02
Plantago schwarzenbergiana	16.79	414.89	46.66
Rorippa kerneri	7.09	632.90	60.65
Trigonella procumbens	3.54	572.70	60.00
Lotus siliquosus	5.97	694.73	86.36
Rhinanthus borbasii ssp. borbasii	66.79	753.30	126.38

Table 2. SOD and GPx activity, GSH content and DPPH-scavenging activity in above-ground parts of selected rare and endangered plant species from the Pannonian plain

Specimen	**SOD activity (U g^{-1} fr. w.)**	**GPx activity (U g^{-1} fr. w.)**	**GSH content (μmol g^{-1} fr. w.)**	**DPPH (% of control)**
Lepidium cartilagineum	642.45	99.37	13.06	8.12
Plantago schwarzenbergiana	776.33	331.25	13.31	46.90
Rorippa kerneri	987.96	248.43	27.03	no activity
Trigonella procumbens	724.39	37.64	30.98	3.99
Lotus siliquosus	844.54	496.87	11.36	7.22
Rhinanthus borbasii ssp. borbasii	139.24	146.14	29.49	24.99

in larger quantities compared to other investigated species. These results were in positive correlation with the amount of MDA in the plant material because this plant species also showed significantly higher MDA quantity (126.38 nmol g^{-1} fr. w.) compared to other samples. According to some authors,[15,16] some rare species of the Pannonian flora, such as *Gallium molugo* L. and *Allium vienale* L., exhibited higher intensity of LP (266.82 and 303.16 nmol g^{-1} f.m., respectively).

One of the reasons why *Rhinantus borbasii ssp. borbasii* specimen showed increased $^{\bullet}O_2^-$ and MDA quantity is probably due to low SOD activity (Table 2). SOD is an inducible enzyme which protects plants from oxidative stress by removing $^{\bullet}O_2^-$ to H_2O_2 and O_2 in dismutation reaction. In *Rhinantus borbasii ssp. borbasii* a disturbance in protective enzyme antioxidant systems could be observed. This has been manifested in strong disproportion between ROS quantity and antioxidant enzymes activities. It is obvious that examined species is incapable to prevent oxidative damage, i.e. it is under oxidative stress. This is supported by the results for lipid peroxidation of plant cell membranes in TBA test. Under such circumstances, the only active antioxidant system in *Rhinantus borbasii ssp. borbasii* was non-enzymatic protective system of glutathione (Table 2). More intensive biosynthesis of this tripeptide is probably due to insufficient activity of other components of antioxidative defence system in plant.

In other investigated specimen, the higher quantities of ROS provoked an antioxidative response from the plants, both in enzymatic and non-enzymatic manner. Besides the increased production of GSH, the activities of antioxidant enzymes SOD and GPx were also high. Compared to the SOD activities in some well-known medicinal and aromatic plants such as *Allium sativum* L. (803.37 U g^{-1} fr. w.), *Achillea millefolium* L. (399.00 U g^{-1} fr. w.) or *Salvia* spp. (125.09–620.54 U g^{-1} fr. w.),[4,16] the SOD activities in leaves of rare Panonnian plants ranged from 139.24 to 988.0 U g^{-1} fr. w. It is well known that the plant species with the higher antioxidant enzymes activities remove ROS more effectively, thus preventing membrane damage and peroxide elimination and degradation to a great extent. Probably the best parameter of the level of oxidative damage in plant organs and tissues is the production of the main end-product of lipid peroxidation — MDA. In our samples, the MDA quantity was low in all investigated species (except in *R. borbasii ssp. borbasii*), which means that plants have not been exposed to significant unfavourable environmental factors.

The content of some non-enzymatic antioxidants such as reduced glutathione (GSH) was also investigated and determined. So-called "small molecules" represent a part of the complex mechanism of tolerance in plants in which they act in reduction of the production of ROS, and in the same time, they scavenge and remove reduced and activated oxygen species that are already present in the cell. It can be seen from our results that, besides in *R. borbasii ssp. borbasii*, the GSH was present in greater quantities also in *Rorippa kerneri* and *Trigonella procumbens*.

The DPPH-radical scavenging activity is a measure of non-enzymatic antioxidant activity. Higher levels of DPPH activity have been correlated with tolerance to different stress conditions.[17] In this assay system, antioxidants can react with the stable free DPPH radical and produce 1,1-diphenyl-2-picrylhydrazine. The change of absorbance produced in this reaction is assessed to evaluate the antioxidant potential of test samples. The EtOH extracts of *Plantago schwarzenbergiana* leaves showed significant levels of the total non-enzymatic antioxidant activity; DPPH-radical scavenging activity was 46.90% higher than control. It seems that plant phenolics and other non-enzymatic antioxidants were active and sufficient to prevent increased MDA formation in this species.

The intensification of agriculture and the quick development of industry about a century ago brought about a continuously growing threat to the plant communities, at all levels of organisation. Habitat loss, different environmental loads resulting in the extinction of some species, invasion by alien species, homogenisation and fragmentation of habitats, pollution and tourism, are among most important threats to all levels of biodiversity. Results obtained showed that the plants were only partially susceptible to the action of ROS. Although present in plant material, ROS could not provoke oxidative stress since quite significant corresponding level of antioxidant defence mechanisms in plants has been established. The only exception to that was *Rhinantus borbasii ssp. borbasii.* Considering high activity of antioxidant enzyme SOD and low LP intensity in most of the plants, it seems that the antioxidant protective systems were sufficient to prevent severe oxidative damage and inactivation of cell components. Thus, it seems that plants were not exposed to significant action of negative environmental stimuli. Also, it seems that their antioxidant status could serve as a suitable biochemical parameter for determination of the level of oxidative stress due to unfavourable environmental factors on a specific location. A special attention should be given to rare and endangered plant species and the monitoring of ROS and antioxidant compounds is proposed.

References

1. Dj. Malenčić, M. Popović, Lj. Nikolić and D. Lazić, Antioxidant status of rare and endangered plants with medicinal properties from the Pannonian part of Serbia, *3rd Conference on Medicinal and Aromatic Plants of Southeast European Countries*, Nitra, Slovak Republic, 22 (2004).
2. S. Stojanović, Dj. Malenčić, D. Lazić, Lj. Nikolić and A. Knežević, *Hottonia palustris L.* — The natural rarity in flora of Vojvodina (taxonomy,

phytogeography and antioxidant status), *International Conference on Sustainable Agriculture and European Integration Processes*, Novi Sad, Serbia, 110 (2004).

3. Dj. Malenčić, Ž. Kevrešan and M. Popović, *Ann. Fac. Eng. Hunedoara.* **3**, 37 (2005).
4. Dj. Malenčić, O. Gašić, M. Popović and B. Pal, *Phytother. Res.* **14**, 546 (2000).
5. O. Blokhina, E. Virolainen and K.V. Fagerstedt, *Ann. Bot.* **91**, 179 (2003).
6. J.G. Scandalios, *Plant Physiol.* **101**, 7 (1993).
7. S. Philosoph-Hadas, S. Meir, B. Akiri and J. Kanner, *J. Agric. Food. Chem.* **42**, 2376 (1994).
8. D. Štajner, R. Igić, B. Popović and Dj. Malenčić, *Phytother. Res.* 2007 (in press).
9. N. Abe, T. Murata and A. Hirota, *Biosci. Biotechol. Biochem.* **62**, 661 (1998).
10. H.D. Misra and I. Fridovich, *J. Biol. Chem.* **247**, 3170 (1972).
11. K.H. Cheesman, A. Beavis and H. Esterbaurer, *Biochem. J.* **252**, 649 (1988).
12. Z.A. Placer, L.L. Cushman and B.C. Johnson, *Anal. Biochem.* **16**, 359 (1966).
13. I. Sedlak, R.H. Lindsay, *Anal. Biochem.* **25**, 192 (1968).
14. L.M. Simon, Z. Fatrai, D.E. Jonos and B. Matkovics, *Biochem. Physiol. Pflanzen* **166**, 387 (1974).
15. D. Štajner, N. Milić and O. Gašić, *Archive of Pharmacy-Belgrade*, **5**, 612 (1997).
16. D. Štajner, N. Milić, N. Mimica-Dukić, B. Lazić and R. Igić, *Phytother. Res.* **12**, 13 (1998).
17. H.M. Kang and E. Saltveit, *Plant Cell Environ.* **25**, 1233 (2002).

CHAPTER 11

HEAVY METALS IN PLANTS — DISTRIBUTION AND METABOLIC EFFECTS

SLAVKO KEVREŠAN, BORIS POPOVIĆ, ĐORĐE MALENČIĆ, MILAN POPOVIĆ, DUBRAVKA ŠTAJNER AND JULIJAN KANDRAČ

Faculty of Agriculture, University of Novi Sad, Novi Sad, Serbia

ŽARKO KEVREŠAN

Food Institute, University of Novi Sad, Novi Sad, Serbia

Distribution of cadmium among different forms (water soluble, Ca-exchangeable and complexed) in the intercellular space and cadmium content in roots, stems and leaves of young sugar beet plants were studied. Nitrate and protein metabolisms of sugar beet plants affected by Ni, Cd and Mo present in concentrations of 10^{-7}, 10^{-5} and 10^{-3} mol dm^{-3} were studied. Nitrate content, activities of nitrate reductase and glutamine synthetase, soluble protein content, and proportion of ribulose-1.5-bisphosphate carboxylase/oxygenase (RuBPCO) protein in leaves of sugar beet plants were determined. The most harmful effect on the biochemical processes in leaves of sugar beet plants had Cd, then Ni, whereas Mo stimulated investigated parameters. The proportion of RuBPCO protein showed a high tolerance to heavy metals.

1. Introduction

Toxic effects of heavy metals (HM) on plant physiology and metabolism are very complex, and they depend on plant species, nature of heavy metal and its concentration. The distinction between the direct and indirect effects of HM action is very difficult. In spite of a large number of studies about the mechanisms of different HM action, further studies are needed, especially at the biochemical level. The aim of this study was to investigate changes in nitrate and protein metabolisms in sugar beet plants caused by Ni, Cd and Mo present in three different concentrations, and Cd distribution in sugar beet plants.

2. Material and Methods

Young sugar beet plants (*Beta vulgaris* L., hybrid NS Hy-11) were grown under semi-controlled greenhouse conditions, in a soil culture using Mitcherlich's pots. Agrochemical characteristics of the soil were: pH in H_2O 8.1 and in KCl 6.7, $CaCO_3$ 2.09%, organic C 2.5%, total N 0.2%, P_2O_5 98 and K_2O 355 mg kg^{-1}(d.m.), presence of the studied HM was not found. Conditions in the greenhouse were as follows: temperature day/night 25/15°C, and 16-h photoperiod. After three weeks, the plants were treated with water solutions of Ni ($NiSO_4$), Cd ($CdCl_2$) and Mo (Na_2MoO_4) in concentrations of 10^{-7}, 10^{-5} and 10^{-3} mol dm^{-3}, the control plants receiving simultaneously the same volume of distilled water.

During the experiment the soil moisture was maintained at 60% of maximal relative soil water capacity, and was constantly monitored gravimetrically. After 12–16 d, i.e., when the symptoms of heavy metal excess appeared under the highest HM concentration, samples of sugar beet leaves for biochemical analyses were taken.

Nitrate content in leaves was determined spectrophotometrically using phenoldisulphonic acid. Nitrate reductase (NR) activity was assayed according to,[1] and glutamine synthetase (GS) activity according to.[2] The RuBPCO protein proportion was estimated as reported earlier.[3] Soluble proteins from sugar beet leaves were extracted with 100 mol m^{-3} phosphate buffer, pH 7, and the protein content was determined after,[4] using bovine serum albumin as standard. Cd distribution in intercellular space was carried out by Kevrešan.[5]

All determinations were made in triplicates, and values were analysed by Duncan's multiple range test and *t*–test.

3. Results and Discussion

Data concerning Cd distribution in sugar beet plants showed that the highest content of Cd was found in roots, and 16% of the total Cd absorbed was transported to the shoot, see Figure 1. These findings are in agreement with[6] who find that Cd transport to the shoots in sugar beet represents 10–20% of total Cd.

Twenty four hours after Cd treatment the highest portion of Cd in intercellular space was found in complexed fraction, see Figure 2. This fraction represents Cd tightly bound to the cell wall.

The applied HM caused a substantial (25% in average) decrease of dry matter content only at the highest concentration (10^{-3} mol dm^{-3}). Of the three metals, the excess of Mo caused the most significant decrease in dry matter content in shoots of sugar beet plants (Table 1). The different effects of a particular HM on plant growth are usually observed at high concentrations; toxic HM concentrations in the majority of cases inhibit plant growth, especially the root growth.[7] Thus, for example, Ni and Cd present in an excess in the substrate caused a significant decrease of dry matter content in maize plants, especially in the roots.[8]

In our study, the nitrate content in the leaves of young sugar beet plants was markedly dependent of both the kind of HM and its concentration (Table 1). With the exception of Mo, an increased concentration of the investigated HM in the substrate caused a significant

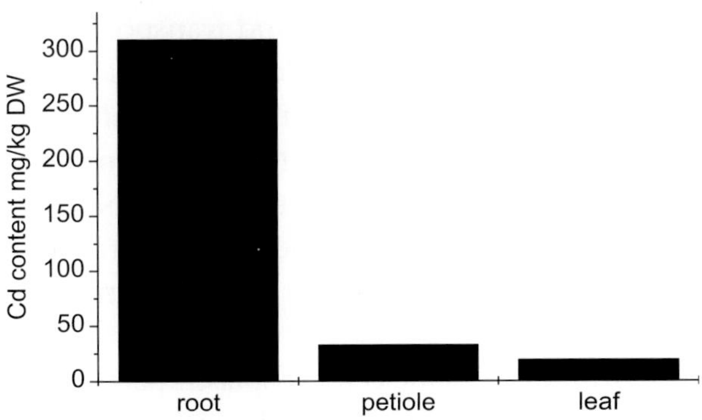

Figure 1. Cadmium content in organs of sugar beet plants.

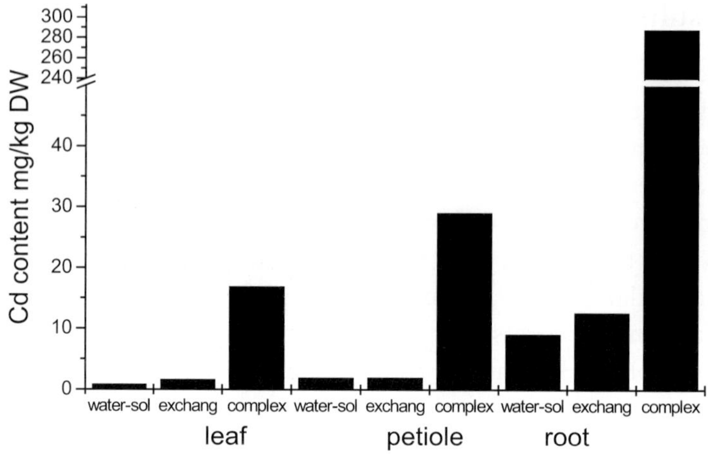

Figure 2. Content of water soluble, exchangeable and complexed cadmium in intracellular space of organs of sugar beet plants.

decrease of nitrate level in plant leaves, especially at the highest HM concentration. The presence of HM in the substrate inhibits uptake of nitrates by the root as well as their translocation by xylem to the site of their reduction. For example, the excess of Cd caused a decrease of nitrate content in shoots of pea plants by 90%.[9] A significant nitrate increase observed in the presence of the highest investigated Mo concentration was probably more a consequence of the pronounced reduction of dry matter content (and thus of the non-existence of a "dilution" effect) than a consequence of a favorable effect of the Mo excess on the uptake and transport of nitrate.

Sugar beet is a plant species in which nitrate is reduced primarily in shoots, whereas the NR activity in the root is insignificant.[10] The measured NR activities were significantly dependent of the nature of the elements and their concentrations (Table 1). Ni caused a significant inhibition of NR only at the highest concentration, whereas Cd reduced substantially this parameter at all concentrations, and especially at the highest one.

The nitrate present in plant cell cytoplasm participates directly in the expression of the NR gene, and thus influences the NR induction.[11,12] Burzynski[13] supposes that HM, and especially Cd, inhibit nitrate translocation from xylem to cytoplasm. The reduced NR activity in photosynthesis tissues is frequently related to the appearance of water deficit in the HM-contaminated plants.[14] Under these conditions, the flux of xylem sap is modified and the rate of transpiration is lowered, i.e. nitrate transport is hindered.[15]

A direct positive correlation was found between the NR activity and the investigated Mo concentrations in the substrate (Table 1). NR and xanthine dehydrogenase are plant enzymes requiring the presence of Mo for their activity. Of the total amount of Mo present in plants, about 10% is built in as the co-factor Mo-pterin in the NR

Table 1. Dry matter content in shoots, and nitrate and soluble protein contents, activities of enzymes of nitrate metabolism, and proportion of ribulose- 1, 5-bisphosphate carboxylase/oxygenase (RuBPCO) in leaves of sugar beet plants Values labeled with * are significantly different from control ($p < 0.05$, t-test), those labeled with ** are significantly different from control ($p < 0.01$, t-test). Values in the same row labeled with different letters are significantly different (Duncan's test, $p < 0.05$)

	Heavy metal concentration (mol dm^{-3})				
	Heavy metal	**Control**	**10^{-7}**	**10^{-5}**	**10^{-3}**
Dry matter	Ni	600.8	$_a$631.5	$_b$531.8	$_b$455.1**
[mg plant^{-1}]	Cd	600.8	$_a$537.8	$_a$519.6	$_a$481.7**
	Mo	600.8	$_a$566.4	$_{ab}$512.2	$_b$417.2**
Nitrate content	Cd	3351	$_a$3338	$_{ab}$2794*	$_b$2501**
[mg kg^{-1}(d.m.)]	Mo	3351	$_a$3382	$_b$2842*	$_b$2455**
	Ni	3351	$_b$2496**	$_b$2650**	$_a$5112**
Nitrate reductase	Ni	0.81	$_a$0.77	$_a$0.65	$_b$0.53**
activity [μmol	Cd	0.81	$_a$0.51**	$_b$0.33**	$_c$0.26**
kg^{-1}(f.m.) s^{-1}]	Mo	0.81	$_a$1.32**	$_a$1.42**	$_b$1.51**
Glutamine synthetase	Ni	14.0	$_a$10.1	$_a$9.3	$_a$6.1**
activity [μmol	Cd	14.0	—	—	—
kg^{-1}(f.m.) s^{-1}]	Mo	14.0	$_c$26.0**	$_b$31.0**	$_a$39.0**
Soluble protein	Ni	27.6	$_a$23.0**	$_b$18.3**	$_b$17.6**
content	Cd	27.6	$_a$23.3*	$_{ab}$21.3**	$_b$19.3**
[g kg^{-1}(f.m.)]	Mo	27.6	$_a$30.6	$_a$32.6**	$_a$33.0**
RuBPCO protein	Ni	26.0	$_a$24.3	$_a$23.8	$_a$22.8
proportion	Cd	26.0	$_a$27.1	$_a$22.9	$_b$18.1*
[% of total proteins]	Mo	26.0	$_a$31.0	$_a$30.7	$_a$26.5

complex.[16] In relation to this, some authors have found that the presence of Mo up to a certain level in the substrate stimulates the NR activity. However, if Mo is present in an excess, then, like other HMs, it can, to a smaller or greater extent, reduce the NR activity.[17]

In literature, there are different opinions on the change of nitrate content and on the activity of particular enzymes of nitrate metabolism in plants in the presence of HM. Yevdokimova[18] found that nitrate is accumulated in plants grown in soils with high contents of HMs. A decrease in nitrate content, accompanied by an increase in the NR activity, was observed in potato plants.[19]

The presence of Cd caused an increase in NR and GS activities of maize leaves,[20] whereas the presence of this element resulted in an insignificant change of GS activity and a reduction of NR activity in pea plants.[21]

The photosynthetic tissues of the majority of cultivated plants contain most often two GS isoenzymes: GS_1 and GS_2.[22] GS_2 may play a primary role in the metabolism of ammonia formed in the chloroplasts after nitrate reduction. With regard to this, the activity of the GS_2 isoenzyme is significantly influenced by NR, nitrite reductase, amount of the substrate, and nitrate content in the cytoplasm.[23] The GS activity in leaves of sugar beet plants was dependent of the nature of the investigated HM and its concentration (Table 1). The applied concentrations of Ni caused a significant decrease in GS activity, whereas the presence of Mo caused its increase. On the other hand, the presence of Cd completely suppressed the GS activity. The excess contents of Cu, Hg, Cd, Zn, Ni, and Fe in the substrate caused a specific inhibition of GS in rice plants.[24] A significant reduction of GS activity in various plant species, and its complete inhibition in sugar beet plants,[25] is sometimes explained by the fact that Cd belongs to the HMs with most pronounced ability to form inactive complexes with the O-, N-, or S-ligands of the structural proteins, especially of the enzyme proteins, inhibiting thus, more or less, their activities.[26] Nevertheless, an excess of Cd inhibits nitrate translocation from the root to the shoot,[9] and the NR activity is mainly influenced by the sum of nitrates present in cell cytoplasm.[13]

The investigated HMs exhibited different effects on the content of soluble proteins in sugar beet leaves. While Ni and Cd caused a decrease of soluble protein content, Mo produced an opposite effect. However, the changes were far less expressed than the changes in activity of the nitrogen assimilation enzymes (Table 1). A high resistance to the HM effect showed also the proportion of RuBPCO protein in sugar beet leaves, significant changes being observed only at the highest Cd concentration (Table 1). Stability of the RuBPCO protein proportion under stress conditions, such as drought, has been observed in sunflower[27] and sugar beet.[28]

Of the investigated HMs the most harmful effect on biochemical processes in sugar beet leaves had Cd, then Ni, whereas the presence of Mo stimulated the investigated parameters. The observed decrease in dry matter content at the highest concentrations of the investigated HMs indicates that the HMs substantially affects various physiological and biochemical processes in plants.[8] This effect has certainly caused a decrease in both the nitrate content in leaves and the NR activity. Because Cd is a mobile chemical element, it is accumulated in leaves, where GS is mostly located in photosynthetic tissues.[29] Hence, there is a positive correlation between Cd concentrations in the xylem and leaves[30] and its direct action on the molecules of NR and GS enzymes can take place, resulting in a pronounced decrease of NR activity and complete inhibition of GS activity. The NR activity can be reduced in the presence of agents for modification of thyol groups of the cysteine moiety,[31] and Cd is known for its specific affinity towards sulphur.[32]

References

1. R.H. Hageman and A.J. Reed, *Method Enzymol.* **69**, 270 (1980).
2. J. Coombs, O.D. Hall, *Techniques in Bioproductivity and Photosynthesis* (Pergamon Press, 1982).
3. S. Kevrešan, N. Petrović, M. Popović and J. Kandrač, *J. Plant Nutr.* **24**, 1633 (2001).
4. O.H. Lowry, N.J. Rosenbrought, A.L.Farr and R.J.Rendall, *J. Biol. Chem.* **193**, 265 (1951).

5. S. Kevrešan, S. Kiršek, J. Kandrač, N. Petrović and Dj. Kelemen, *Biol. Plantarum.* **46**, 85 (2003).
6. S. Lindberg and G. *Physiol. Plantarum.* **63**, 181 (1985).
7. S. Karataglis, M. Moustakas and L. Symeonidis, *Biol Plant.* **33**, 3 (1991).
8. H.W. Woolhouse, Toxicity and tolerance in the response of plants to metals, *Encyclopedia of Plant Physiology.* Vol. 12C, O.L. Lange, P.S. Nobel, C.B. Osmond, H. Ziegler, Ed., Springer Verlag, 246 (1983).
9. E.L. Hernández, R. Carpena-Ruiz and A. Gárate, *J. Plant Nutr.* **19**, 1581 (1996).
10. N. Petrović and R. Kastori, Nitrate reductase in sugar beet genotypes supplied with different nitrate levels, *Genetic Aspects of Plant Mineral Nutrition*, N. El Bassam, *et al.*, Eds. 51–55 (Kluwer Academic Publishers, 1990).
11. T. Hoff, M.B. Stummann and W.K. Henninsen, *Physiol. Plant.* **84**, 616 (1992).
12. R. Tischner, B. Waldeck, S.P. Goyal and W.D. Rains, *Planta.* **189**, 533 (1993).
13. M. Burzynski, *Acta. Soc. Bot. Pol.* **57**, 349 (1988).
14. W.E. Rauser and E.B. Dumbroff, *Env. Exp. Bot.* **21**, 249 (1981).
15. J. Barcelo and C. Poschenrieder, *J. Plant. Nutr.* **13**, 1 (1990).
16. B.A. Notton, *Annu. Proc. Phytochem. Soc. Eur.* **21**, 219 (1983).
17. H.S. Srivastava, *Phytochemistry* **19**, 725 (1980).
18. G.A. Yevdokimova, *Eurasian Soil. Sci.* **26**, 116 (1994).
19. N.I. Mondy and C.B.Munshi, *J. Agr. Food. Chem.* **41**, 256 (1993).
20. M. Ferretti, R. Ghisi, L. Merlo, F. Dallavecchia and C. Passera, *Photosynthetica* **29**, 49 (1993).
21. L.K. Chugh, V.K. Gupta, S.K. and Sawhney, *Phytochemistry* **31**, 395 (1992).
22. F.S. Mc Nally, R. Hirel, A.H. Gadol, A.F. Mann and R.G. Stewart, *Plant Physiol.* **72**, 22 (1983).
23. A. Oaks, I. Stulen, K. Jones, J.M. Winspear, S. Mishra and L.I. Boesel, *Planta.* **148**, 477 (1980).
24. B.J. Miflin and P.J. Lea, Ammonia asasimilation, *The Biochemistry of plants,* J.B. Miflin, Ed. 169 (Academic Press, 1980).
25. M. Popović, S. Kevrešan, J. Kandrač, J. Nikolić, N. Petrović and R. Kastori, *Biol. Plant.* **38**, 281 (1996).
26. F. Van Assche and H. Clijsters, *Plant Cell Environ.* **13**, 195 (1990).

27. D Panković, Z. Sakač, S. Kevrešan and M. Plesničar, *J. Exp. Bot.* **50**, 127 (1999).
28. S. Kevrešan, B. Krstić, M. Popović, L. Kovačev, S. Pajević, J. Kandrač and Đ. Malenčić, *Biol. Plant.* **40**, 245 (1997/98).
29. F. Dubois, N. Brugiüre, R.S. Sangwan and B. Hirel, *Plant Mol. Biol.* **31**, 803 (1996).
30. Y.T. Guo and H. Marschner, *J. Plant Nutr.* **18**, 2691 (1995).
31. A.J. Trimboli, G.B. Quinn, E.T. Smith and M.J. Barber, *Arch. Biochem. Biophys.* **331**, 117 (1996).
32. W. Kaim and B. Schwederski, *Bioinorganic Chemistry: Inorganic Elements in the Chemistry of Life* (John Wiley and Sons, 1994).

CHAPTER 12

EFFECT OF SHORT-TERM WATER DEFICIENCY ON VASCULAR TISSUE OF PETIOLE AND LEAF LAMINA OF SUGAR BEET

MILORAD PERČIĆ AND IVANA MAKSIMOVIĆ

Faculty of Agriculture, University of Novi Sad, Novi Sad, Serbia

LANA KRSTIĆ AND JADRANKA LUKOVIĆ

Faculty of Sciences, University of Novi Sad, Novi Sad, Serbia

Anatomical changes in vascular tissues of different sugar beet genotypes, induced by water deficiency, are discussed in this paper. The total cross-section area of phloem, xylem and average cross-section area of vessels of petiole and lamina were measured in ten genotypes exposed to short-term water deficiency. In seven genotypes all measured parameters increased, while in three of them they decreased compared to control plants. Statistically significant differences were found in G5 and G6 (increase in all parameters) and G8 and G11 (decrease in all parameters). These results suggest that there is a difference in the effect of short-term water deficiency on vascular tissues of petiole and lamina in different genotypes.

1. Introduction

Different abiotic and biotic factors are limiting for reaching the genetic yield potential of crop plants.[1] One of the most common abiotic stress factors is drought.

Sugar beet (*Beta vulgaris, ssp. vulgaris*, L.) produces large vegetative mass, and therefore, has high water requirements. Water deficiency is very high during summer, due to insufficient amount of precipitation.[2] Depending on the intensity and length of drought yield losses can vary from 30% to 50%.[3]

In order to define drought tolerance in sugar beet genotypes various parameters of plant development and functioning must be tested.[4] Water deficiency in plants may induce different physiological, biochemical and molecular changes. Those changes include closure of stomata, decrease of photosynthesis intensity, reduced leaf area and finally, reduced organic matter production.[5,6]

The aim of this paper was to investigate anatomical changes in vascular tissues of different sugar beet genotypes, induced by short-term water deficiency, and to point out genotypes that are more tolerant to drought and therefore suitable for further selection for drought tolerance improvement.

2. Materials and Methods

Ten genotypes of sugarbeet (*Beta vulgaris* ssp. *vulgaris*), marked as G1–G6, G8–G9 and G11–G12, were included in the analyses. In preliminary experiments, under field conditions, these genotypes have shown significant differences in drought tolerance. Plants were grown in semi-controlled conditions, in pots with soil/sand mixture in a glasshouse, and watered daily to 80% of FWC. Ninety days after sowing water was withheld completely for five days to induce short-term water deficiency.

Twelve microscopical slides (six for control and six for water deficiency group), each containing 3 cross-sections were made for each genotype using cryostat apparatus Leica CM 1850, at the temperature from 18 to 20°C, with cutting intervals from 20 to 30 μm. Computer software *Image Analysing System Motic* was used for measuring selected anatomical parameters.

The total cross-section area of phloem, xylem and average cross-section area of vessels of petiole and lamina were measured and compared between stressed (group subjected to short-term water deficit, marked as WD) and watered (marked as Control) plants of each genotype.

The results were statistically processed by analysis of variance followed by *t*-test between control and stressed plants of each genotype using software *Statistica 7*. Statistically significant differences are marked with * at $\alpha_{0.05}$ and with ** at $\alpha_{0.01}$ in tables and figures.

3. Results and Discussion

Sugar beet genotypes responded differently to experimental conditions. In seven genotypes all measured parameters increased, while in three they decreased compared to the control plants. Statistically significant differences were mostly found in G5 and G6 (increase in all parameters) and G8 and G11 (decrease in all parameters) (Tables 1 and 2).

Even though the measured parameters differ greatly in control plants (Control) and plants exposed to water deficiency (WD), those differences are often not statistically significant (Tables 1 and 2). The reason is that analysed parameters have shown high standard deviation in all tested genotypes.

The total cross-section area of phloem and xylem of lamina main vein, see Figure 1, were significantly higher in G5 and G6 while they were significantly lower in G8 and G11. An average cross-section area of vessels lumen was significantly higher only in G5 while it was significantly lower in G8 and G11. Values for plants exposed to water deficiency (WD) in figures are shown as a percentage of the control (Control = 100%).

Vascular tissue of petiole has shown significant changes in more genotypes than vascular tissue of lamina, see Figure 2. The total cross-section area of phloem in vascular bundles of petiole significantly increased in three genotypes (G3, G4 and G6) while it decreased in two genotypes (G8 and G11). The total cross-section area of xylem significantly increased in G2, G6 and G12 while it significantly decreased in G8 and G11. An average cross-section area of vessels significantly increased in G5, G6 and G12, while it significantly decreased in G11.

Table 1. Vascular tissue of the lamina main vein in sugar beet genotypes (data in table represent mean value, standard deviation (SD ±) and coefficient of variation (CV)). The values with superscripts * and ** differ significantly for $\alpha_{0.05}$ and $\alpha_{0.01}$ respectively

	Vascular tissue of the lamina main vein					
	Total phloem cross-section area (mm^2)		Total xylem cross-section area (mm^2)		Average cross-section area of vessels lumen (μm^2)	
Genotype	Control	WD	Control	WD	Control	WD
G1	0.023 SD ± 0.011 CV 47.3	0.029 SD ± 0.017 CV 59.1	0.028 SD ± 0.019 CV 69.3	0.034 SD ± 0.020 CV 60.0	221.3 SD ± 74.5 CV 33.7	237.5 SD ± 65.0 CV 27.4
G2	0.036 SD ± 0.008 CV 21.1	0.034 SD ± 0.012 CV 37.6	0.038 SD ± 0.012 CV 30.8	0.037 SD ± 0.013 CV 33.2	353.9 SD ± 89.3 CV 25.2	276.1 SD ± 113.1 CV 39.1
G3	0.025 SD ± 0.009 CV 35.2	0.016 SD ± 0.006 CV 35.9	0.029 SD ± 0.013 CV 45.2	0.022 SD ± 0.010 CV 46.2	304.5 SD ± 129.9 CV 42.6	234.8 SD ± 45.0 CV 19.2
G4	0.019 SD ± 0.008 CV 38.9	0.027 SD ± 0.007 CV 24.2	0.027 SD ± 0.011 CV 39.8	0.034 SD ± 0.011 CV 33.2	320.8 SD ± 133.3 CV 41.6	283.5 SD ± 53.3 CV 18.8
G5	0.020 SD ± 0.007 CV 37.3	0.035** SD ± 0.008 CV 24.2	0.036 SD ± 0.017 CV 45.6	0.070** SD ± 0.020 CV 28.5	330.2 SD ± 96.4 CV 29.2	444.4* SD ± 93.8 CV 21.1

(*Continued*)

Table 1. (*Continued*)

Genotype	Vascular tissue of the lamina main vein: Total phloem cross-section area (mm^2)		Total xylem cross-section area (mm^2)		Average cross-section area of vessels lumen (μm^2)	
	Control	WD	Control	WD	Control	WD
G6	0.020	0.044**	0.023	0.061*	287.2	394.1
	SD ± 0.008	SD ± 0.015	SD ± 0.010	SD ± 0.029	SD ± 82.1	SD ± 137.1
	CV 40.9	CV 33.2	CV 44.3	CV 47.5	CV 28.6	CV 34.8
G8	0.065	0.016**	0.119	0.020**	409.3	246.6**
	SD ± 0.022	SD ± 0.005	SD ± 0.030	SD ± 0.006	SD ± 68.5	SD ± 40.2
	CV 34.4	CV 31.6	CV 25.6	CV 30.8	CV 16.7	CV 16.3
G9	0.039	0.038	0.059	0.071	361.8	337.8
	SD ± 0.020	SD ± 0.009	SD ± 0.036	SD ± 0.027	SD ± 73.6	SD ± 98.4
	CV 50.4	CV 24.2	CV 60.0	CV 38.0	CV 20.4	CV 29.1
G11	0.098	0.052**	0.220	0.125**	553.1	470.3
	SD ± 0.020	SD ± 0.008	SD ± 0.041	SD ± 0.017	SD ± 97.3	SD ± 133.3
	CV 20.8	CV 15.0	CV 20.1	CV 13.7	CV 17.6	CV 28.4
G12	0.040	0.042	0.060	0.061	418.7	383.5
	SD ± 0.022	SD ± 0.014	SD ± 0.037	SD ± 0.019	SD ± 59.5	SD ± 68.0
	CV 54.4	CV 33.6	CV 62.8	CV 30.7	CV 14.2	CV 17.7

Table 2. Petiole vascular tissue of sugar beet genotypes (data in table represent mean value, standard deviation (SD ±) and coefficient of variation (CV)). The values with superscripts * and ** differ significantly for $\alpha_{0.05}$ and $\alpha_{0.01}$ respectively

	Petiole vascular bundles					
	Total phloem cross-section area (mm²)		Total xylem cross-section area (mm²)		Average cross-section area of vessels lumen (μm²)	
Genotype	Control	WD	Control	WD	Control	WD
G1	0.078 SD ± 0.054 CV 68.4	0.056 SD ± 0.026 CV 46.5	0.106 SD ± 0.081 CV 76.3	0.073 SD ± 0.035 CV 48.1	440.6 SD ± 114.3 CV 25.9	449.4 SD ± 85.8 CV 19.1
G2	0.101 SD ± 0.033 CV 32.9	0.102 SD ± 0.028 CV 27.5	0.122 SD ± 0.027 CV 22.2	0.171* SD ± 0.045 CV 26.4	576.5 SD ± 125.1 CV 21.7	624.3 SD ± 161.5 CV 25.9
G3	0.111 SD ± 0.066 CV 59.9	0.121* SD ± 0.022 CV 18.5	0.111 SD ± 0.043 CV 38.6	0.165 SD ± 0.042 CV 25.4	506.6 SD ± 130.1 CV 25.7	619.2 SD ± 97.2 CV 15.7
G4	0.077 SD ± 0.015 CV 20.2	0.104* SD ± 0.021 CV 20.0	0.113 SD ± 0.036 CV 31.8	0.118 SD ± 0.024 CV 20.0	347.7 SD ± 56.6 CV 16.3	399.5 SD ± 50.3 CV 12.6
G5	0.088 SD ± 0.040 CV 45.2	0.124 SD ± 0.013 CV 10.6	0.160 SD ± 0.107 CV 66.8	0.218 SD ± 0.050 CV 22.9	667.9 SD ± 163.7 CV 24.5	1057.9* SD ± 355.9 CV 33.6

(*Continued*)

Table 2. *(Continued)*

Genotype	Petiole vascular bundles: Total phloem cross-section area (mm^2)		Total xylem cross-section area (mm^2)		Average cross-section area of vessels lumen (μm^2)	
	Control	WD	Control	WD	Control	WD
G6	0.089	0.130**	0.107	0.208**	413.5	744.9*
	SD ± 0.015	SD ± 0.022	SD ± 0.023	SD ± 0.044	SD ± 73.3	SD ± 198.1
	CV 17.1	CV 17.3	CV 1.9	CV 21.2	CV 17.7	CV 26.6
G8	0.185	0.053**	0.329	0.077**	432.7	347.7
	SD ± 0.070	SD ± 0.004	SD ± 0.152	SD ± 0.011	SD ± 111.4	SD ± 56.6
	CV 37.9	CV 8.5	CV 46.2	CV 14.0	CV 25.8	CV 16.3
G9	0.087	0.097	0.180	0.229	523.9	630.7
	SD ± 0.032	SD ± 0.027	SD ± 0.090	SD ± 0.071	SD ± 160.2	SD ± 213.5
	CV 37.2	CV 28.3	CV 50.1	CV 31.0	CV 30.6	CV 33.8
G11	0.183	0.110**	0.435	0.282*	869.6	697.7*
	SD ± 0.042	SD ± 0.030	SD ± 0.123	SD ± 0.110	SD ± 127.4	SD ± 103.8
	CV 22.9	CV 27.7	CV 28.2	CV 38.9	CV 14.6	CV 14.9
G12	0.087	0.104	0.142	0.202*	426.2	650.4*
	SD ± 0.020	SD ± 0.025	SD ± 0.034	SD ± 0.055	SD ± 128.2	SD ± 134.1
	CV 22.8	CV 24.4	CV 23.8	CV 27.1	CV 30.1	CV 20.6

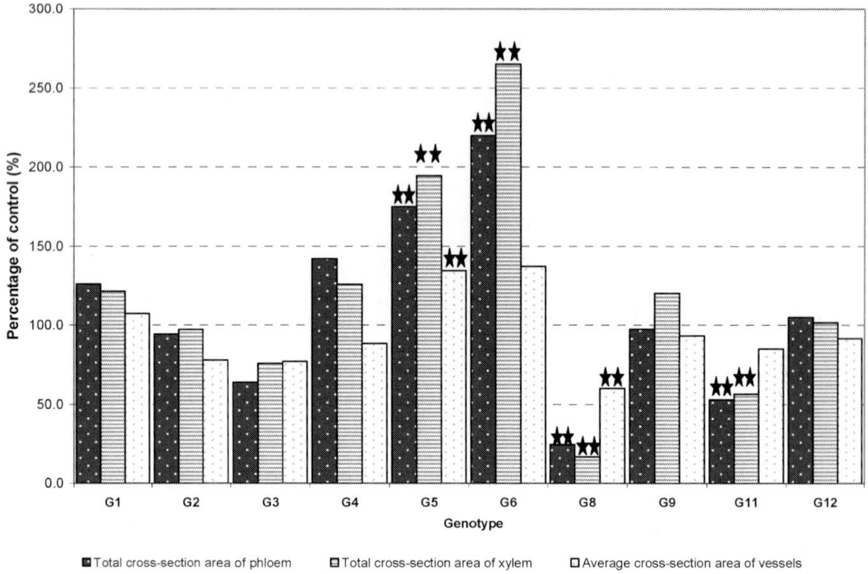

Figure 1. Vascular tissue of lamina main vein in sugar beet genotypes.

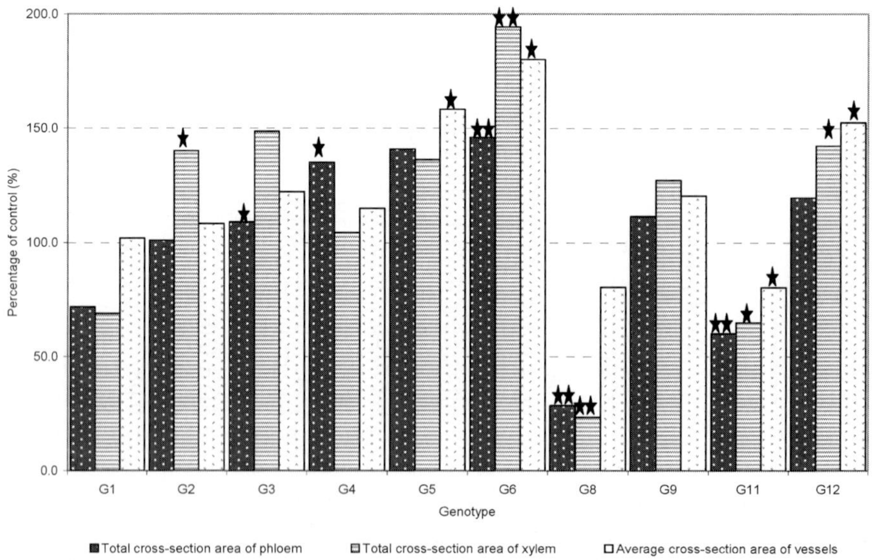

Figure 2. Petiole vascular tissue of sugar beet genotypes.

4. Conclusion

In this paper changes in vascular tissue of lamina and petiole of ten different genotypes of sugar beet were tested under drought conditions. These are the conclusions that can be made:

1. More statistically significant changes were found in petiole vascular tissue then in vascular tissue of lamina.
2. Four genotypes showed statistically significant changes in lamina main vein vascular tissue (G5 and G6 reacted with increase, and G8 and G11 with decrease of total cross-section area of phloem and xylem). Only G8 reacted with significant decrease of average cross-section area of vessels lumen.
3. In eight out of ten genotypes significant changes in tested parameters of vascular bundles of petiole were found. Total cross-section area of phloem significantly increased in G3, G4 and G6 while it significantly decreased in G8 and G11. Total cross-section area of xylem significantly increased in G2, G6 and G12 and decreased in G8 and G11. Average cross-section area of vessels statistically significantly increased in G5, G6 and G12 while it significantly decreased only in G11.

Genotypes in which vascular tissue of lamina and petiole were less altered under experimental conditions may be considered as more tolerant to water deficiency. G1 and G9 were not significantly altered, and in G5 and G6 statistically significant increase of vascular tissue of lamina and petiole were found under drought. These facts should be taken into consideration during next process of selection.

References

1. C. Itai, A. Nejidat and D. Bar Zvi, Stress physiology in new millenium, *Plant Physiology in New Millenium,* S.A. Quarie, B. Krstić, V. Janjić Eds., Yugoslav Society of Plant Physiology, 31 (2002).
2. Ž. Popović, R. Stikić, *Agrochemistry* **1**, 53 (1986).
3. K.W. Jaggard, A.M. Dewar and J.D. Pidgeon, *J. Agr. Sci.* **130**, 337 (1998).

4. M.M. Chaves, J.P. Maroco and J.S. Pereir, *Funct. Plant. Biol.* **30**, 239 (2003).
5. J. Flexas, J. Bota, G. Loreto and T.D. Sharkey, *Ann. Appl. Bio.* **144**, 273 (2004).
6. M. Perčić, I. Maksimović, N. Nagl and L. Kovačev, The effect of water deficiency on some characteristics of sugar beet leaves, *7th International Symposium Young People and Multidisciplinary Research*, Resita, Romania, EB 33 (2005).

CHAPTER 13

THE EFFECT OF NITRATE CONCENTRATIONS ON PHYSIOLOGY OF DIFFERENT LEAF AGE GROUPS

ANDREJ PILIPOVIĆ, SAŠA ORLOVIĆ AND MARINA KATANIĆ

Institute of Lowland Forestry and Environment, University of Novi Sad, Novi Sad, Serbia

MILAN BORIŠEV AND NATAŠA NIKOLIĆ

Faculty of Sciences, University of Novi Sad, Novi Sad, Serbia

The aim of this research was to investigate the effect of increased nitrate concentrations in modified Hoagland solution on physiology of different leaf age groups as one of parameters for selection of genotypes for nitrate phytoremediation. Poplar cuttings were taken from three following clones of genus Populus: (1) Populus x euramericana cl. Pannonia, (2) Populus deltoides cl. B-81 and (3) hybrid between P. nigra x maximowitzzi and P. nigra var. Italica cl. 9111/93. During the plant growth, besides the biomass production of plants and leaf area (LA), physiology of different leaf age groups (LPI) was investigated through measuring parameters related to nitrate assimilation such as nitrate reductase activity (NRA), nitrate accumulation in leaf age groups and total nitrogen content in leaves and roots. The results showed different reaction of poplar clones to the given nitrate concentrations, where clone B-81 (Populus deltiodes) showed the best results and stability of the investigated parameters under different treatments, which indicates its ability for nitrate phytoremediation.

1. Introduction

Phytoremediation is an emerging technology that uses various plants to degrade, extract, contain or immobilise contaminants from soil and water.[1] Among the various number of plant species, only few wooden species are suitable for phytoremediation. Poplars (*Populus sp.*) and willows (*Salix sp.*) are the most common tree species used for phytoremediation because they grow rapidly, have many and deep roots, and take up large quantities of water,[2] as shown in numerous scientific papers.[3–8]

Among the various numbers of contaminants present in the environment, nitrates play significant role in ground and surface water contamination. According to,[9] if nitrates present in drinking water are converted to nitrites they can cause methemoglobinemia in infants, while excess nitrate in surface waters, together with phosphorus, can cause eutrophication. This contamination is mostly caused by agricultural practices, where fertilization with the increased application of mineral fertilizers plays a significant role in increase of nitrate concentrations both in plant tissues and in the environment.[10] Also, excess of nitrates in soil promotes their denitrification and emission of nitrous oxides, such as N_2O, thus causing even up to 300 times more harmful effects than CO_2.[11]

Considering the importance and share of agriculture in Serbia, especially in the northern province of Vojvodina, and long tradition of poplar growing and researches conducted at the Institute of Lowland Forestry and Environment, the aim of this study was to investigate potentials of different poplar clones for phytoremediation and obtain information about physiological parameters to be used in future selection of poplar genotypes.

2. Material and Methods

Cuttings from four poplar genotypes, which represent a wider spectrum of genus *Populus*: *Populus x euramericana "Pannonia"*, *Populus deltoides cl. B-81* and hybrid (*P. nigra x maximowctzzi) x P. nigra* "Italica" *cl. 9111/93* (inter-section hybrid), were taken in early spring of 2004. Nine cuttings from each genotype were placed in 36 hydroponic pots of 40 litres in volume and grown in pure water for 45 days, until the beginning of root development. Plants were grown in semi-controlled greenhouse at temperature of 21 ±1°C and day-night photoperiod. After root initiation, plants were treated to grow under three different concentrations of nitrates: (i) in modified Hoagland solution with 2 mM NO_3 concentration, (ii) standard Hoagland nutrient solution with 10 mM of NO_3 concentration, and (iii) in modified Hoagland solution with 30 mM NO_3 concentration. Nutrient solutions were changed once per week, and constantly aerated.

During the growing period, the nitrate reductase activity (NRA) of different leaf age groups (young/mature) according to Leaf Plastochron Index (LPI) classification,[12] and Nitrate Reductase Activity (NRA) in roots was assessed. For NRA determination leaf with LPI = 2 value (NRA_2) was taken as representative young leaf, while mature leaf had LPI = 7 value (NRA_7). At the end of 135 days growth period, plants were removed and their leaf area (LA) measured. After drying of plant material in oven, total biomass of plants (M) measured in g_{DW}, and nitrogen assimilation and nitrate accumulation was assessed in leaves and roots. Nitrate accumulation was determined in two different leaf age groups of LPI < 7 (young leaves) and LPI > 7 (mature leaves). Nitrate reductase activity, nitrate assimilation of roots and shoots were analysed according to and samples were read out on spectrophotometer (Cecil CE 2021). Nitrogen assimilation in dried plant material was determined also according to the same author. Calculation was done using Microsoft Office EXCEL and MSTAT C programme by Duncan test.

3. Results

Biomass production of investigated clones increased with the increase of nitrate concentrations in nutrient solution, see Table 1. The difference in biomass (M) among genotypes was not significant only between clones 9111/93 and "Pannonia" grown at concentration of 10 mM NO_3 in nutrient solution. The effect of treatment showed differences in all clones, but 2 mM NO_3 and 10 mM NO_3 treatments

Table 1. The effect of nitrate concentration on biomass production (M), leaf area (LA); *values with same letter did not differ significantly at p 0.05 regarding each parameter separately

	Concentration of NO_3 in the growth medium					
Clone	**2 mM**		**10 mM**		**30 mM**	
	M(g)	*LA(cm²)*	*M(g)*	*LA(cm²)*	*M(g)*	*LA(cm²)*
9111/93	12.9e	768e*	15.5d	1010d	19.5c	1389c
Pannonia	15.4d	889de	17.5cd	1087d	24.3b	1488c
B-81	18.2c	1639c	25.6b	2341b	44.9a	3993a
Average	15.5c	1099c	19.5b	1501b	29.6a	2290a

did not cause statistically significant differences in biomass of clone "Pannonia". Highest M of plants grown on 30 mM NO_3 had clone B-81 (44.93 g_{DW}).

The results of measurement of leaf area (LA) in Table 1 showed also increase due to elevated nitrates in the nutrient solution. Clone B-81 differed significantly in all treatments, in relation to other two clones, while treatment of 10 mM NO_3 did not increase significantly leaf area of clone "Pannonia", compared to treatment of 2 mM NO_3. Results also showed superiority of clone B-81 in vigor where leaf area of investigated clone in 2 mM NO_3 treatment was greater than leaf area of clones 9111/93 and "Pannonia" in all treatments, although in some cases difference was not significant.

The NRA of young leaves (LPI = 2) shown in Table 2 decreased for all clones submitted to 2 and 30 mM NO_3 treatments, although differences in NRA activity in clone "Pannonia" young leaves were not significant. Mature leaves were less sensitive to given treatments with insignificant differences between them, but the NRA was more intensive than in young leaves. Regardless of clone, treatments did not cause significant differences in NRA of both young and mature leaves.

Nitrate accumulation in leaves (Figure 1) increased with the elevation of nitrates in nutrient solutions both for clones and leaf ages. Young leaves showed greater accumulation than mature leaves in all cases except clone B-81 in the highest treatment, where young leaves

Table 2. The effect of nitrate concentration on leaf nitrate reductase activity (NRA) of investigated clones; NRA_2 — nitrate reductase activity of young leaves (LPI = 2) in μmol NO_2 g^{-1} h^{-1}; NRA_7 — nitrate reductase activity of mature leaves (LPI = 7) in μmol NO_2 g^{-1} h^{-1}

	Concentration of NO_3 in the growth medium					
Clone	**2 mM**		**10 mM**		**30 mM**	
	NRA_2	*NRA_7*	*NRA_2*	*NRA_7*	*NRA_2*	*NRA_7*
9111/93	1.638b	2.683a	2.313a	1.933a	1.605b	3.474a
Pannonia	1.268b	2.566a	1.250b	2.955a	1.798ab	3.631a
B-81	1.380b	2.172a	2.303a	2.581a	1.150b	2.826a
Average	1.429a	2.474a	1.955a	2.489a	1.518a	3.310a

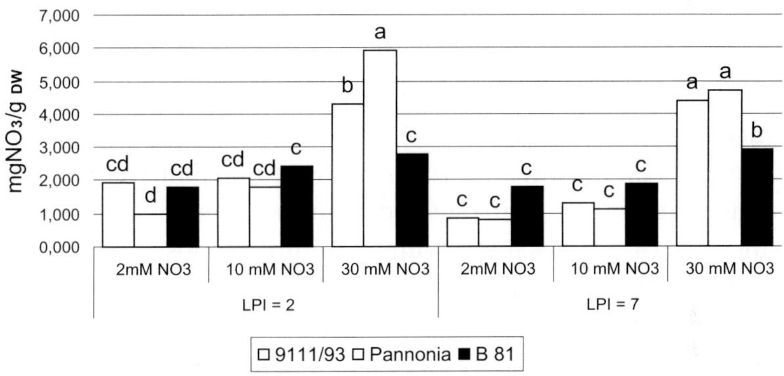

Figure 1. Nitrate accumulation in young (LPI = 2) and mature (LPI = 7) leaves.

accumulated more nitrates than mature ones. The highest accumulation of nitrates was assessed in young and mature leaves of clone "Pannonia" plants grown under 30 mM NO_3 treatment.

Clone B-81 showed no significant differences in nitrate accumulation in young leaves between different treatments. Accumulation in older leaves was also enhanced by increasing nitrate concentration in solutions. Significant differences were recorded only between treatments, with exception of clone B-81 in 30 mM NO_3 treatment, where accumulation was lowered in relation to 9111/93 and "Pannonia" clones.

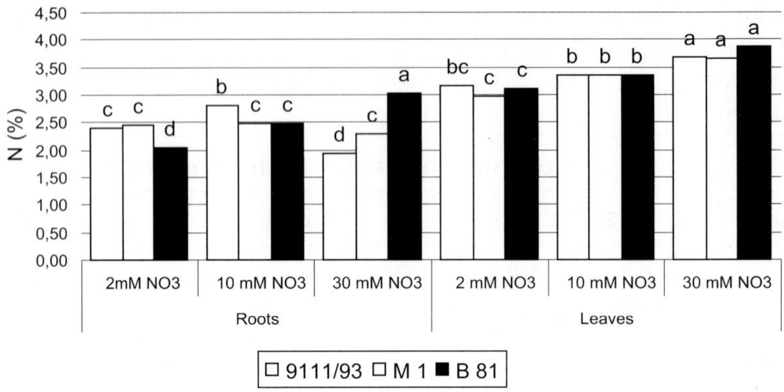

Figure 2. Nitrogen content in leaves and roots of clones grown under different concentrations of nitrates (%).

Nitrogen content in leaves (Figure 2) did not differ significantly among clones under the same treatment, while its differences between treatments were significant.

Contents of nitrogen in roots showed variability both between clones and treatments. Nitrogen content in roots of clone B-81 grew with increase of nitrates in growing medium. Opposite to clone B-81, clones 9111/93 and "Pannonia" had decreased amount of nitrogen in roots of plants grown under 30 mM NO_3 treatment. The highest nitrogen content in roots of 9111/93 was assessed in 10 mM NO_3 treatment. Generally, nitrogen content of investigated clones was higher in leaves than in roots.

4. Discussion

The results obtained for biomass production and leaf area demonstrated significant influence of nitrate concentrations in nutrient solution. Biomass of plants grown at higher amount of nitrates was greater than of those grown at lower amount of available nitrates. Similar results are recorded in researches conducted with hybrid poplar[13,14] and aspen,[15] which showed strong effect of plant available nitrogen to growth parameters of plants. Besides the effect of nitrogen concentration on growth parameters, clonal variability was also expressed. Differences in growth parameters of various clones were thoroughly analysed and discussed in many studies[16–20] where differences between both genotypes of same and different poplar species, including their hybrids, were recorded. Investigated growth parameters of clones "Pannonia" and 9111/93 showed similar results probably due to their genetic closure originated from the European black poplar (*P. nigra*).

Nitrates assimilation process is closely related with the amount of nitrates in the substrate and NRA.[19] Enzymatic activity increases proportionally with increase of nitrate concentration in substrate until a point is reached from which NRA remains constant,[20,21] or declines with further increase of nitrates.[22] Assessment of NRA in this research showed increase from lower (2 mM) to medium (10 mM) concentration, while high amount (30 mM) decreased NRA in young leaves (LPI = 2) but remained constant in mature leaves (LPI = 7).

Nitrate accumulation showed correlation with the amount of nitrate available to the plants. The concentration of plant tissue increased with increased of nitrates in nutrient solution. Similar research was conducted on wheat,[23] beech and ash,[24] aspen[25] and poplar.[26] From phytoremediation viewpoint the role of roots is to store nitrates from soil and groundwater and serve as their reservoir.[13] Yet another evidence substantiating this presumption is the fact that nitrate accumulation in roots had value that was almost equal to total nitrogen amount in roots.

The total nitrogen amount in plant tissues is also dependent upon available amounts in the substrate. Treatments of plants with different dosages either of fertilizers[27,28] or municipal wastewaters containing significant amounts of nitrogen[15] showed increase of total nitrogen in plant tissues.

The results of this research paper showed different reactions of poplar clones to given nitrate concentrations, with clone B-81 (*Populus deltiodes*) showing best results and stability of investigated parameters under different treatments, thus indicating its ability for nitrate phytoremediation. In addition, results also demonstrated possibility of determining new parameters in poplar breeding such as potential for phytoremediation.

References

1. Introduction to Phytoremediation, U.S. Environmental Protection Agency, National Risk Management Research Laboratory Office of Research and Development Cincinnati, Ohio, EPA/600/R-99/107 (2000).
2. L.A. Licht and J.G. Isebrands, *Biomass and Bioenerg.* **28**, 203 (2005).
3. Z. Galić, A. Pilipović, B. Klašnja, S. Orlović and V. Vasić, *Poplar* **72**, 177 (2006).
4. M. Xingmao and J. Burken, *Environ. Sci. Technol.* **38**, 4580 (2004).
5. L.A. Licht, J.L. Schnoor, Tree buffers protect shallow ground water at contaminated sites, EPA Ground Water Currents, Office of Solid Waste and Emergency Response, EPA/542/N-93/011 (1993).
6. M. Katanić, Z. Tomović, A. Pilipović, S. Orlović and B. Krstić, *Poplar* **88**, 177 (2006).
7. G.S. Bañuelos, H.A. Ajwa, B. Mackey, L.L. Wu, C. Cook, S. Akohoue and S. Zambrzuski, *J. Environ. Qualit.* **26**, 639 (1997).
8. A. Pilipović, N. Nikolić, S. Orlović, N. Petrović and B. Krstić, *Z. Naturforsch.* **60c**, 247 (2005).
9. P. Newbould, *Plant. Soil.* **115**, 297 (1989).
10. R. Kastori and N. Petrović, *Nitrates in Vegetables: Physiological, Ecological and Agrotechnical Aspects*, Institute of Field and Vegetable Crops, Novi Sad, Verzal (2003).
11. J.S. Schepers, D.D. Francis and J.F. Shanahan, *J. Bioscience.* **60**, 186 (2005).
12. D.I. Dickmann, *Bot. Gaz.* **132**, 252 (1971).
13. Gr.J. O'Neill and A.M. Gordon, *J. Environ. Qualit.* **23**, 1218 (1994).
14. J.E.K. Cooke, T.A. Martin and J.M. Davis, *New Phytol.* **167**, 41 (2005).
15. J.B.J. Hart and P.V. Nguyen, *J. Environ. Qualit.* **23**, 1218 (1994).
16. R. Ceulemans, Genetic variation in functional and structural productivity components in *Populus*: Causes and consequences of variation, *Growth Rate and Productivity of Higher Plants*, Eds. H. Lambers *et al.*, SPB Academic Publishing, 69 (1989).
17. S. Orlović, B. Klašnja and P. Ivanišević, *Poplar* **72**, 159 (1997).
18. T.S. Barigah, B. Saugier, M. Mosseau, J. Guittet and R. Ceulemans, *Ann. Forest. Sci.* **51**, 613 (1994).
19. S. Orlović, V. Guzina, B. Krstić and Lj. Merkulov, *Silvvae. Genet.* **47**, 183 (1998).

20. F. Devienne-Barret, E. Justes, J.M. Machet and B. Mary, *Ann. Bot.* **86**, 995 (2000).
21. N. Shankar and H.S. Srivastava, *Biol. Plantarum* **39**, 583 (1997).
22. B.M. Chen, Z.H. Wang, S.X. Li, G.X. Wang, H.X. Song and X.N. Wang, *Plant. Sci.* **167**, 635 (2004).
23. C.X. Hu, B.E. Deng and T.C. Liu, *J. Huazhong Agric. Univ.* **11**, 239 (1992).
24. M.O. Vouillot, J.M. Machet and J.M. Meynard, *Eur. J. Agron.* **5**, 227 (1996).
25. G. Gebauer, E.D. Schulze, Nitrate nutrition of Central European forest trees, *Trees — Contributions to Modern Tree Physiology*, H. Rennenberg, W. Eschrich, H. Ziegler Eds. 273 (Backhuys Publishers, 1997).
26. X.M. Min, Y. Siddiqi, R.D. Guy, A.D.M. Glass and H.J. Kronzucker, *Plant. Cell. Environ.* **21**, 1039 (1998).
27. B.L. Black, L.H. Fuchigami and G.D. Coleman, *Tree Physiol.* **22**, 717 (2002).
28. L.J. Samuelson, *New Forest.* **19**, 95 (2000).

CHAPTER 14

IMPROVING BRADYRHIZOBIUM — SOYBEAN SYMBIOSIS

NASTASIJA MRKOVAČKI AND MILICA HRUSTIĆ

Institute of Field and Vegetable Crops, Novi Sad, Serbia

The objective of this study was to assess the effects of Nitragin application, Nitragin and PGPR (plant growth promoting rhizobacteria) — Azotobacter and Nitragin re-inoculation 30 days after first inoculation of soybean. The trial with different inoculation methods was established at Rimski Šančevi experiment field in 2001 and 2002. The soybean variety Afrodita (created in the Institute of Field and Vegetable Crops in Novi Sad) was used in experiments. The trial included the following variants: 1. control — non-inoculated variant (Ø); 2. soybean seeds inoculated with NS–Nitragin (microbiological fertilizer for soybean) before sowing (N); 3. soybean seeds inoculated with NS–Nitragin and sprayed with Azotobacter suspension before sowing (N + A); 4. soybean seeds inoculated with NS–Nitragin during sowing and 30 days later by simultaneous incorporation of mixture of strains for NS–Nitragin in the soil (N + N). On average for the two years, the respective variants increased the yield of soybean variety Afrodita by 15%, 17.6% and 20.4% in relation to non-inoculated variants.

1. Introduction

Dinitrogen fixation provides more N to agricultural ecosystems worldwide than the total amount of N fertilizers applied. The significant potential still exists to further improve the symbiotic and associative N_2 fixation by breeding genotypes with a greater capacity to sustain these interactions with bacteria.

Better N_2-fixing symbiosis may be achieved both by manipulating rhizobia and plant hosts and by eventually creating an artificial rhizosphere. The improvements in symbiotic relationships between rhizobia and legumes may result from the introduction of the third partner.[11] The N_2 fixation was enhanced by AM combination in comparison with non-mycorhizal alfalfa plants.[13] Co-inoculation in the field with *Rhizobium* and *Azospirillum* promoted root development, nodulation, nitrogen fixation and crop yield in a wide variety of legumes.[15] Many studies suggest that there are good possibilities for increasing nodulation, nitrogen fixation and crop yield of legumes by inoculation with *Rhizobium* and *Azospirillum* or *Azotobacter*.[7,8,12,14]

The objective of this study was to assess the effects of Nitragin application, Nitragin and PGPR (plant growth promoting rhizobacteria) — *Azotobacter* and Nitragin re-inoculation 30 days after first inoculation of soybean.

2. Material and Methods

The trial with different inoculation methods was established at Rimski Šančevi experiment field in 2001 and 2002. The soybean variety Afrodita (created in the Institute of Field and Vegetable Crops in Novi Sad) was used in experiments. The trial included the following variants: 1. control — non-inoculated variant (Ø); 2. soybean seeds inoculated with NS–Nitragin (microbiological fertilizer for soybean) before sowing (N); 3. soybean seeds inoculated with NS–Nitragin and sprayed with *Azotobacter* suspension before sowing (N + A); 4. soybean seeds inoculated with NS–Nitragin during sowing and 30 days later by simultaneous incorporation of mixture of strains for NS–Nitragin in the soil (N + N).

The bacterial strains for NS–Nitragin (patent 416/95, Mrkovački *et al.*, 1995) were cultivated in Demolon's substrate at the temperature of 30°C combined with aeration for 72 hours. The density of the strains was 10^9 per 1 ml of liquid culture. *Azotobacter chroococcum* strains from NSCNFB (Novi Sad Collection Nitrogen Fixing Bacteria registered in World Data Centre on Microorganisms in Japan) were cultivated in Fiodor's substrate in a shaker at 30°C for 48 hours. The density of the strains was 10^9 per 1 ml of liquid culture.

The four-replicate trial was set up in mid-May. The elementary plot size was 2 × 5 m per variant. Plant samples were taken two and four months after sowing, i.e., in July and September, respectively. The samples taken in July were analysed for the number of nodules, plant weight and nitrogen contents in the aboveground part, root and the whole plant. The samples taken in September were analysed for grain parameters, pod parameters, aboveground part weight and grain yield.

3. Results and Discussion

The 2001 growing season was rainy. It was one of the most favourable years for soybean production in last three decades. The average soybean yield for the Vojvodina Province was 2.55 t/ha. The weather conditions in 2002 were specific. Water deficit occurred in the first part of the growing season, but it turned out that the rainfall in the period July-August was decisive for yield formation. The average soybean yield for the Vojvodina Province, including the location of Rimski Šančevi, was between 2.0 and 3.0 t/ha.

In 2001, all three types of inoculation increased the mass and N contents in roots, aboveground plant parts and nodules. The number of nodules was significantly increased and the mass of nodules was highly significantly increased in the variants N and N + A (Table 1).

At the end of the 2001 growing season, the number of grains per plant, grain mass per plant and N content in grain were increased while the yield was highly significant in relation to that in the non-inoculated variant (Table 2).

In 2002, all three types of inoculation brought increases in plant mass and N content and highly significant increases in the number of nodules. Also, there was a highly significant increase in N content in the aboveground plant part (Table 3).

At the end of the 2002 growing season, significant and highly significant increases were registered in the number of grains (N + A),

Table 1. Effect of different types of inoculation on soybean mass, nitrogen content and number of nodules at the stage of flower (2001)

	Mass (g/Plant)				N content (mg/Plant)				No. of nodules/ Plant
	Shoot	Root	Nodules	Whole Plant	Shoot	Root	Nodules	Whole Plant	
Ø	4.79	0.73	0.07	5.59	132.3	8.20	4.15	144.68	3.66
N	6.79	1.08	0.26	8.13	196.7	12.36	14.39	223.85	16.33
N + A	5.65	0.88	0.23	6.76	161.5	11.41	19.51	185.45	18.56
N + N	4.85	0.75	0.19	5.78	147.2	9.87	10.13	167.25	15.56
LSD									
5%	2.22	0.57	0.10	2.79	57.19	4.76	6.72	63.98	5.44
1%	3.36	0.86	0.15	4.22	86.64	7.21	10.17	96.93	8.25

Table 2. Effect of different types of inoculation on soybean yield per ha, grain number and mass and N content in grain (2001)

	Yield t/ha	No. of grains/Plant	Grain mass/Plant	N content mg
Ø	2350	75.89	10.73	599.10
N	2846	88.78	13.32	813.52
N + A	2926	77.45	15.73	982.76
N + N	3053	102.11	15.46	915.96
LSD				
5%	302.43	15.17	3.28	162.88
1%	458.15	22.98	4.96	246.75

grain mass (N, N + A) and N content in grains (N, N + A). Yields were significantly higher in all three types of inoculation (N, N + A, N + N) as compared with the non-inoculated variant (Table 4).

The highest grain yield for both years was obtained in the variant of re-inoculation (N + N), followed by the combination N + A (Tables 2 and 4). On two-year average, the variants N, N + A and N + N increased the grain yield by 15%, 17.6% and 20.4%, respectively, in relation to the non-inoculated control. The respective increases obtained in an earlier study[10] were 17%, 19% and 25%. The efficacy of inoculation varies depending on several factors, all of which affect the

Table 3. Effect of different types of inoculation on soybean mass, nitrogen content and number of nodules at the stage of flower (2002)

	Mass (g/Plant)				N content (mg/Plant)				No. of nodules/ Plant
	Shoot	Root	Nodules	Whole Plant	Shoot	Root	Nodules	Whole Plant	
Ø	4.60	0.71	0.03	5.36	144.39	6.81	1.73	152.92	2.56
N	8.65	1.19	0.07	9.89	240.88	15.51	3.58	259.97	16.11
N + A	7.01	1.05	0.06	8.12	239.03	12.73	3.47	255.22	15.67
N + N	8.51	1.42	0.08	9.01	252.80	19.06	4.39	276.25	18.33
LSD									
5%	3.70	0.32	0.03	3.62	40.26	5.62	2.05	102.60	9.35
1%	5.61	0.48	0.05	5.49	61.00	8.52	3.10	155.43	14.6

Table 4. Effect of different types of inoculation on soybean yield per ha, grain number and mass and N content in grain (2002)

	Yield t/ha	No. of grain/Plant	Grain mass/Plant	N content mg
Ø	2853	48.33	7.63	444.34
N	3030	69.56	12.24	721.17
N + A	3073	77.56	13.48	818.92
N + N	3093	58.33	10.08	598.66
LSD				
5%	228.46	15.9	2.21	131.71
1%	346.09	24.09	3.35	199.53

number of viable rhizobia available for infection of legume roots. Increased numbers of viable rhizobia per seed by application of inoculant above the commercially recommended rate, results in a continued linear increase in nodulation and yield. Several studies have reported yield increases up to 25%.[5]

Studies carried out by Ciafardini and Barbieri (1987) obtained a higher rate of nitrogen fixation with the repeated inoculation 30 days after sowing than with a single inoculation at sowing, Burns *et al.* found that mixed cultures of *Azotobacter vinelandii* and several *Rhizobium* species caused increases in the number of root nodules on their respective hosts as compared with plants inoculated with *Rhizobium* alone.

Milić *et al.* (2001, 2002) conducted a field trial with soybean and concluded that the repeated inoculation (N + A) improved plant growth and grain yield. Cvijanović (2002) achieved the highest yields of soybean with the application of *Azotobacter* and NS-Nitragin.

Three hypotheses have been proposed to account for the reported increases in plant growth subsequent to inoculation with *Azotobacter*.[1] These are: the provision of combined nitrogen to the plant via nitrogen fixation by *Azotobacter*, changes in the rhizosphere microflora that make conditions more conducive to plant growth and the production of plant growth substances by *Azotobacter*. The last hypothesis has been accepted as generally sufficient to explain the inoculation responses although the suppression of plant pathogens by *Azotobacter* may also be important in some cases.[2]

The plant growth substances are known to exert activity at extremely low concentrations and have been demonstrated to be present in supernatants of *Azotobacter* cultures.[3,4,6,9]

References

1. F.E. Allison, *Soil Sci.* **64**, 413 (1947).
2. M.E. Brown, *Annu. Rev. Phytopatol.* **12**, 181 (1974).
3. M.E. Brown, S.K. Burlingham, *J. Gen. Microbiol.* **53**, 135 (1968).
4. M.E. Brown and N. Walker, *Plant Soil.* **32**, 250 (1970).
5. R. Deaker, R.J. Rougley and I.R. Kennedy, *Soil Biol. Biochem.* **36**, 1275 (2004).
6. M. Govedarica, V. Milić and Đ. Gvozdenović, *Plant Soil.* **42**, 113 (1993).
7. E.E. Iruthayathas, S. Gunasekaran and K. Vlassak, *Sci Hortic-Amsterdam.* **20**, 231 (1983).
8. Y.V.D.K. Kumar Rao and R.B. Patil, *Curr. Sci. India.* **45**, 523 (1976).
9. N. Mrkovački and V .Milić, *Ann. of Microbiol.* **51**, 145 (2001).
10. N. Mrkovački and M. Hrustić, *J. Sci. Agricult. Res.* **64**, 165 (2003).
11. Y. Okon, S. Burdman, M. German, R. Itzigsohn, S. Sarig, S. Vanderleyden, J.T.E.M. Odemdaal, P. Dutto and C. Labandera–Gonzalez, The potential for cereal and legume crop yield promotion by Azospirillum

under sub — Optimal growth conditions, *3rd European Nitrogen Fixation Conference*, Lunteren, The Netherlands, 66 (1998).

12. J. Plazinski, E. Gartner, R. Jahnke and B.G. Rolfe, Effect of Azospirillum strains on Rhizobium legume symbiosis, *Advances in Nitrogen Fixation Research*, C. Vegger and W. E. Newton Eds., 424 (1984).
13. Z. Rengel, *Plant Soil.* **245**, 147 (2002).
14. C.S. Singh and N.S. Subba Rao, *Plant Soil.* **53**, 387 (1979).
15. M. Toro, R. Azcon and J.M. Barea, *New Phytol.* **138**, 273 (1998).
16. N. Mrkovacki, V. Milic, Z. Saric, Patent 416/95, Selection of high effective strains of Bradyrhizobium Japonicum. *Beograd* (1995).
17. G. Ciafardim and C. Barbieri, *Agrou. J.* **79**, 645 (1987).
18. V. Milic, N. Mrkovacki, M. Hrustic, *Eurosoya.* **12**, 20 (2001).
19. V. Milic, N. Mrkovacki, M. Popovic, D. Malencic, *Rostlinna Vyroba.* **48**, 356 (2002).
20. G. Cvijanovic, Effect of diazotrophs on yield and microbiological acitivity in soil under maize, wheat and soyabean. PhD theses Faculty of Agriculture, University of Novisad (2002).

CHAPTER 15

CYTOLOGICAL CHARACTERISTICS OF REGENERANTS OBTAINED FROM ANTHER CULTURE OF WHEAT (*TRITICUM AESTIVUM* L.)

BRANKA LJEVNAIĆ, ANKICA KONDIĆ-ŠPIKA, BORISLAV KOBILJSKI AND NIKOLA HRISTOV

Institute of Field and Vegetable Crops, Novi Sad, Serbia

The aim of this study was to investigate ploidy levels of regenerated green plants obtained from anther culture of wheat (Triticum aestivum L.). Anthers were isolated from 20 F_1 hybrid combinations and grown on Potato-2 inductive medium. Calluses were transferred to the 190–2 regeneration medium on which 145 green plants were obtained. The cytological analysis showed that, out of the total number of green plants, 42.8% were haploids ($2n = 3x = 21$) and 53.8% were spontaneous double haploids ($2n = 6x = 42$). Only 3.4% regenerated green plants were aneuploids ($2n \neq 21, 42$).

1. Introduction

Green plants regenerated in *in vitro* wheat anther culture mostly have the haploid number of chromosomes — *haploids* ($2n = 3x = 21$) or the diploid number of chromosomes — *double haploids* ($2n = 6x = 42$). Among haploid and double haploid plants, the appearance of *aneuploids* ($2n \neq 21, 42$) is also possible.

Haploid plants are completely sterile because all their chromosomes are univalent. But, their haploid callus tissues are extremely valuable for many genetic[1,2] and physiological studies[3,4] as well as for genetic engineering.[5]

Double haploids in wheat anther culture may occur naturally by spontaneous chromosome doubling. When they occur from haploid cells, they are completely homozygous for all allelic genes. These spontaneous double haploids are used in wheat breeding programs as pure lines.[6–8]

The aim of this study was to investigate ploidy levels of regenerated green plants obtained from anther culture of 20 randomly selected F_1 wheat combinations.

2. Materials and Methods

Twenty F_1 hybrid combinations of wheat (*Triticum aestivum* L.) were used for anther isolation. They represented a part of selection material of Small Grains Department, Institute of Field and Vegetable Crops in Novi Sad. From each hybrid combination, 600 anthers were isolated. Isolation, sterilisation and growing of anthers and green plants were carried out according to the standard procedure described by.[9]

Before green plants transfer to pots, the top of the root was taken from each plant for cytological analysis. Fixing of material and acetocarmine staining were carried out according to the standard procedure described by.[10] After that, cytological analysis was conducted and the percentages of haploids, spontaneous double haploids and aneuploids determined.

3. Results and Discussion

Cytological analysis of green regenerants acquired by wheat anther culture showed that most of the used hybrid combinations produced haploid and spontaneous double haploids. From the total number of regenerated green plants (145 green plants), 42.8% (or 62 plants) were haploids $/2n = 3x = 21/$, see Figure 1 and 53.8% (or 78 plants) were spontaneous double haploids $/2n = 6\times = 42/$, see Figure 2. Only 5 or 3.4% of the regenerated green plants were aneuploids$/2n \neq 21, 42/$, see Figure 3 whereof 2.1% were hexaaneuploids ($2n = 40$,

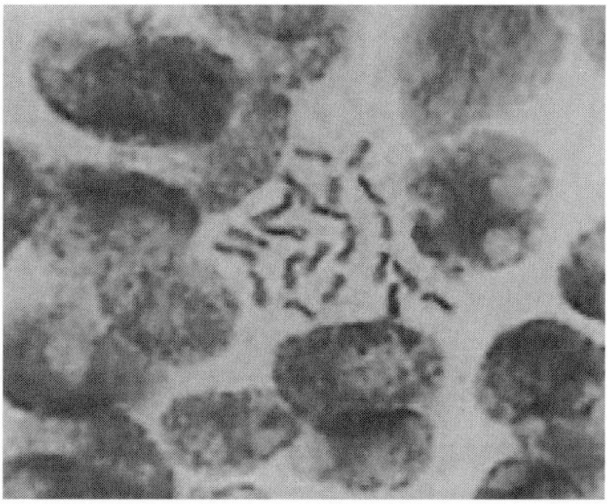

Figure 1. The chromosome number in a haploid wheat plant.

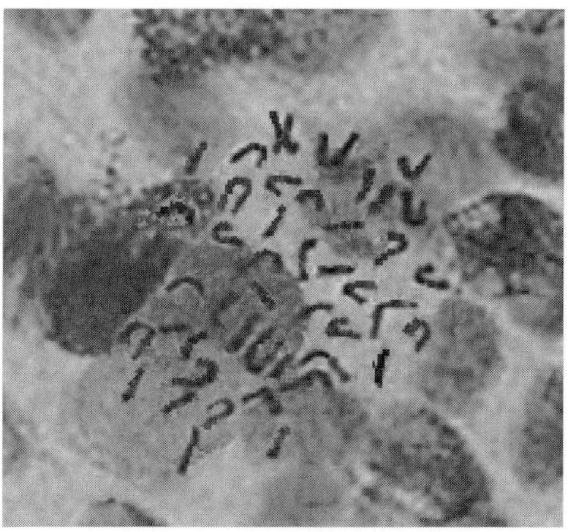

Figure 2. The chromosome number in a hexaploid wheat plant.

41, 43 or 44) and 1.3% were haploaneuploids (2n = 19, 20 or 21). Euploids (2n = 8× = 56, 2n = 9x = 63) were not found. These results are similar to the results obtained by.[9,11]

Hybrid combination Banks × Chris regenerated all haploid green plants, while hybrid combinations NS 109-96 × Radika, Radika × Pesma and Prima × Huequen produced only spontaneous double

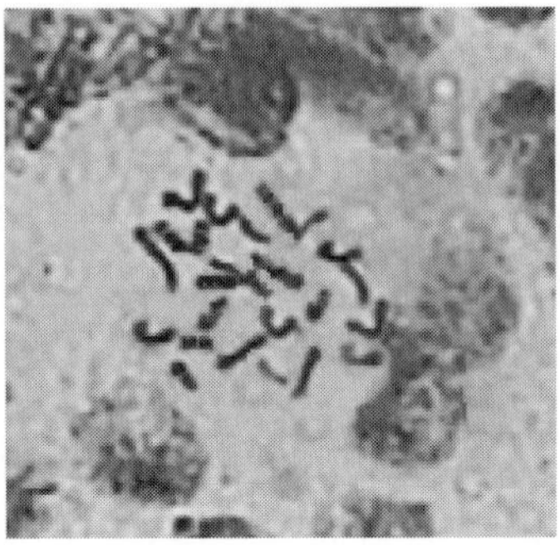

Figure 3. The chromosome number in an aneuploid wheat plant.

Table 1. Percentages of haploids, spontaneous double haploids and aneuploids obtained by wheat anther culture

Hybrid combination	Num. of green plants	Haploids		Spontaneous double haploids		Aneuploids	
		Num	%	Num	%	Num	%
Prima × Huequen	3.0	1.0	33.3	2.0	66.7	0.0	0.0
NS 173-98 × NS 164-98	4.0	1.0	25.0	3.0	75.0	0.0	0.0
Mexico-3 × Radika	13.0	3.0	23.1	10.0	76.9	0.0	0.0
Radika × Huequen	3.0	2.0	66.7	1.0	33.3	0.0	0.0
Prima × Garazinko	9.0	3.0	33.3	5.0	55.6	1.0	11.1
NS 163-98 × NS 121-98	4.0	1.0	25.0	3.0	75.0	0.0	0.0
NS 163-98 × NS 164-98	5.0	2.0	40.0	2.0	40.0	1.0	20.0
NS 163-98 × NS 178-98	4.0	1.0	25.0	2.0	50.0	1.0	25.0
NS 109-96 × Radika	3.0	0.0	0.0	3.0	100.0	0.0	0.0
NS 163-98 × Radika	3.0	1.0	33.3	2.0	66.7	0.0	0.0
Radika × Pesma	3.0	0.0	0.0	3.0	100.0	0.0	0.0
Anastazija × Radika	4.0	2.0	50.0	2.0	50.0	0.0	0.0
Banks × Chris	4.0	4.0	100.0	0.0	0.0	0.0	0.0
Banks × Pinzon Inta	5.0	2.0	40.0	3.0	60.0	0.0	0.0
Prima × Bezostaja-1	6.0	2.0	33.3	4.0	66.7	0.0	0.0
Banks × F 53-70	33.0	17.0	51.5	16.0	48.5	0.0	0.0
NS 42-00 × Sofija	13.0	7.0	53.9	4.0	30.8	2.0	15.4
NS 30-95 × NSP-11	22.0	13.0	59.1	9.0	40.9	0.0	0.0
Banks × Pobeda	0.0	0.0	0.0	0.0	0.0	0.0	0.0
Prima × Huequen	4.0	0.0	0.0	4.0	100.0	0.0	0.0
Percentage	7.3	3.1	42.8	3.9	53.8	0.3	3.4

haploids. Hybrid combination Banks × Pobeda showed no green plant regeneration ability. Except for these extreme values, the percentages of the obtained haploid plants ranged between 23.1% (Mexico-3 × Radika) and 66.7% (Radika × Huequen), while the percentages of spontaneous double haploids were in an interval from 30.8% in NS 42-00 × Sofija to the 76.9% in hybrid combination Mexico-3 × Radika (Table 1).

Hybrid combination NS 42-00 × Sofija regenerated two haploaneuploid plants too, while hybrid combinations Prima × Garazinko; NS 163-98 × NS 164-98 and NS 163-98 × NS 178-98 each regenerated one hexaaneuploid plant (Table 1).

4. Conclusion

Out of the total number of green plants regenerated by wheat anther culture, 42.8% were haploids, 53.8% were spontaneous double haploids and 3.4% were aneuploids (2.1% were hexaaneuploids and 1.3% were haploaneuploids). Euploids were not found.

Combination Banks × Chris produced only haploid green plants, while combinations NS 109-96 × Radika, Radika × Pesma and Prima × Huequen produced only spontaneous double haploids. Combination NS 42-00 × Sofija regenerated two haploaneuploid plants, while hybrids Prima × Garazinko, NS 163-98 × NS 164-98 and NS 163-98 × NS 178-98 each produced one hexaaneuploid plant.

References

1. K. Borojević, *Conntemp. Agric.* **33**, 371 (1985).
2. A.M. Torp, A.L. Hansen and S.B. Andersen, *Euphytica* **119**, 377 (2001).
3. O.A. Monastyrsky, *Russian Agric. Sci.* **2**, 1 (1999).
4. M.V. Salomon, C.E. Camargo de Oliveira and A.W.P. Ferreira Filho, *Bragantia* **62**, 189 (2003).
5. H.E. Lörz, E. Gobel and P. Brown, *Plant. Breed.* **100**, 1 (1988).
6. S. Šesek and S. Denčić, Doubled haploids in wheat breeding, *Proc. of 1st Balkan Symp. Breeding and Cultivation in Wheat, Sunflower and Legume Crops*, Albania, Bulgaria, (1995).

7. S. Tuvesson, A. Ljungberg, N. Johansson, K.E. Karlsson, L.W. Suijs and J.P. Josset, *Plant. Breed.* **119**, 455 (2000).
8. I. Zamani, E. Gouli-Vavdinoudi and D.G. Roupakias *http://www.cab-subsets.org/* (1990).
9. B. Ljevnaić, A. Kondić-Špika, B. Kobiljski and S. Denčić, *Genetics* **38**, 153 (2006).
10. S. Petrović, and M. Vučenović, *Condinium of Citogenetics*, Faculty of Agriculture, University of Novi Sad, Novi Sad (1992).
11. K.M. Kim and S.P. Baenziger, *In Vitro Cell. Dev. B.* **41**, 22 (2005).

CHAPTER 16

TRAITS RELATED WITH TOLERANCE TO EXCESS BORON IN WHEAT

MILKA BRDAR AND BORISLAV KOBILJSKI

Institute of Field and Vegetable Crops, Novi Sad, Serbia

IVANA MAKSIMOVIĆ AND MARIJA KRALJEVIĆ-BALALIĆ

Faculty of Agriculture, University of Novi Sad, Novi Sad, Serbia

Excess boron can limit the plant growth and yield, particularly in arid and semiarid environments. Sixty wheat genotypes were included in the trial in order to investigate the relationship among traits related to tolerance to excess boron and to determine the most informative treatment. Root length, dry weight and boron content were measured at seedling stage. Genotypes with small root growth suppression on treatments, namely boron tolerant genotypes, generally had smaller dry weight reduction and lower boron content. Positive correlations among diverse boron treatments were found for all studied traits; therefore, any analysed treatment may be informative enough for further research.

1. Introduction

Boron is an essential plant micronutrient, yet high levels of soil boron can seriously affect the growth and yield of many crop species, particularly in arid and semiarid environments. Excess boron occurs naturally in association with saline soils, such as solonchaks and solonetz, which are common in Vojvodina province. High concentrations of soil boron can also be related to industrial chemicals, fly ash and surface mining. Because soil amelioration is not an economic proposition, the best approach to solve problems associated with boron toxicity is breeding and cultivation of tolerant genotypes.[10]

Differences in susceptibility to excess boron exist at both intra- and inter-specific levels.[3] Great variability among bread and durum wheat have been reported by many authors.[4,5,11] However, there are several impediments associated to diagnosing tolerance to excess boron. The range between boron toxicity and deficiency is narrower than that for any other element.[2] It is very difficult to predict the growth of plants on boron laden soils, because availability of boron to plants depends on various physical and chemical properties of the soil.[14] Foliar and shoot analyses are also associated with considerable problems, which are mainly due to the steep gradient of boron within leaves, the effect of environmental conditions on boron accumulation and the leaching of boron from leaves by rain.[9]

Y. Chantachume, D. Smith, G. J. Hollamby, J. G. Paull and A. J. Rathjen[1] suggested a filter paper technique as cheap, rapid and non-destructive test for assessing response of wheat to excess boron. The purposes of this study were to investigate the relationships among traits related to tolerance to excess boron in wheat (*Triticum aestivum* L.) using the filter paper technique and to determine the most informative treatment for further research.

2. Material and Methods

The experimental material consisted of 60 divergent wheat (*Triticum aestivum* L.) genotypes. Cultivars and advanced lines created at the Institute of Field and Vegetable Crops in Novi Sad (Republic of Serbia), high-yielding foreign cultivars and cultivars with known reaction on high levels of boron have been included in the trial. The trial was conducted at the Small Grains Department of the Institute of Field and Vegetable Crops and the filter paper technique[1] was used. We surface sterilised 1200 kernels of each genotype and conducted pre-germination on 4°C for 48 h and on 18°C for 24 h. The kernels were germinated and the seedlings have grown on filter paper soaked with different (0, 50, 100 and 150 mg/l) concentrations of boric acid (H_3BO_3) solution (B 0-control; B 50, B 100 and B 150-treatments) at 18°C, for 11 days. All solutions included 0.5 mM $Ca(NO_3)_2$ x $4H_2O$, 0.0025 mM $ZnSO_4 \times 7H_2O$ and 0.015 mM H_3BO_3.

The samples included 50 seedlings. Root length (RL-cm), dry weight (W-mg) and boron content (BC-mg/kg dry matter) were measured. Relative root length (RRL-%), relative dry weight (RW-%) and relative boron content (RBC) on treatments were calculated with regard to control (control = 100%). Numerical values corresponding to RBC were transformed and they represented the ratio of boron content on treatments and on control. Boron content was determined by ICP-spectrophotometer and samples were prepared by method 4, suggested by.[7]

Relative root length (RRL) was used as selection criterion for boron tolerance. Genotypes were classified into four groups: boron tolerant (T-RRL above 90%), medium tolerant (MT-RRL 90–80%), medium sensitive (MS-RRL 80–70%) and sensitive (S-RRL lower than 70%). Mean, minimum and maximum values of the analysed traits were calculated for all groups. Spearman's coefficients of rang correlation (r) were calculated for all pairs of treatments in order to determine the most informative boron treatment. Spearman's coefficients of rang correlation were also calculated for pairs of traits on control, as well as on treatments, in order to investigate their relationships. STATISTICA 7.0 software package[15] was used.

3. Results and Discussion

The RRL was used as a selection criterion for boron tolerance in wheat and analysed genotypes were classified in four groups. Fourteen genotypes were marked as boron sensitive (S), seventeen as medium sensitive (MS) and nineteen as medium tolerant (MT). Ten studied genotypes might be considered as boron tolerant (T). All analysed traits varied in vast intervals within groups (Table 1).

Vast intervals of variation regarding RRL, RW and RBC may confirm the proposal given by[5] that there are more than two regions in the wheat genome involved in boron tolerance. They found that the region on chromosome 7B was associated with the control of boron uptake, root growth suppression and leaf symptom expression.

Table 1. Mean, minimum and maximum values of traits related to boron tolerance in seedlings of 60 wheat genotypes

	Group	RL	RRL	W	RW	BC	RBC	No
Mean	T	30.4	+4.0	11.0	+1.3	14.2	13.8	10
Min		21.4	90.7	6.9	87.7	5.4	4.5	
Max		37.5	+61.4	15.1	+17.4	23.4	32.0	
Mean	MT	35.4	84.5	11.1	97.6	17.7	9.5	19
Min		13.0	81.3	6.0	84.8	9.7	5.6	
Max		53.7	88.7	15.1	+10.0	26.6	18.7	
Mean	MS	37.7	74.1	10.8	91.6	15.3	13.8	17
Min		23.3	70.2	8.6	77.2	3.4	7.0	
Max		72.6	79.0	13.7	+19.2	27.3	51.1	
Mean	S	31.8	61.9	8.9	86.6	12.9	17.0	14
Min		16.2	47.4	5.3	66.5	2.8	10.3	
Max		43.1	69.2	11.7	+4.9	19.6	69.5	

RL, W, BC-root length (cm), dry weight (mg) and boron content (mg/kg dry matter) in wheat seedlings on control; RRL, RW, RBC-relative root length (%), dry weight (%) and boron content (transformed data) in wheat seedlings on boron treatments with regard to control; T, MT, MS, S-groups of boron tolerant, medium tolerant, medium sensitive and sensitive wheat genotypes; No-number of genotypes in group.

The region on chromosome 7D was also associated with leaf symptom expression. J. G. Paull[12] proposed that regions on 4A and 7D chromosomes were related to boron toxicity tolerance in wheat. Exceedingly vast intervals of variation concerning RBC within groups of genotypes, especially within sensitive and tolerant group, imply the possibility of different boron tolerance mechanisms[6,13] also proposed different mechanisms of boron tolerance.

However, we found significantly negative correlations between RRL and RBC and between WR and RBC (Table 3). This suggests that mechanism of boron tolerance in genotypes included in this study is generally based on avoidance of boron uptake.

Spearman's coefficients of rang correlation between control and boron treatments with respect to root length of wheat seedlings were insignificant (Table 2), which implies that differences in RRL are not caused by inherent differences in root length among the

Table 2. Spearman's coefficients of rang correlation (r) for root length, dry weight and boron content in seedlings of 60 wheat genotypes on control (B0) and boron treatments (B50, B100, B150)

	Control-treatment		
Pairs	**r (RL)**	**r (W)**	**r (BC)**
B0/B50	–0.19 ns	–0.05 ns	–0.83**
B0/B100	0.03 ns	–0.02 ns	–0.86**
B0/B150	–0.01 ns	0.12 ns	–0.85**
	Treatment-treatment		
Pairs	**r (RRL)**	**r (RW)**	**r (RBC)**
B50/B100	0.51**	0.55**	0.91**
B50/B150	0.46**	0.41**	0.89**
B100/B150	0.63**	0.63**	0.91**

RL, W, BC-root length, dry weight and boron content in wheat seedlings on control; RRL, RW, RBC-relative root length, dry weight and boron content in wheat seedlings on boron treatments with regard to control. ns, **-insignificant, significant at the 0.01 level of probability, respectively.

studied genotypes and that root growth suppression in the presence of excess boron could be used as selection criterion for boron tolerance.

Correlation coefficients between the control and treatments regarding seedling's dry weight were also insignificant (Table 2). However, root length and dry weight of seedlings were positively correlated both on control and boron treatments (Table 3). Differences in seedling's dry weight may be due to differences in root length and WR alone should not be considered as selection criterion for boron tolerance in wheat.

Negative correlation coefficients between control and treatments were found for seedling's boron content (Table 2), which is in disagreement with the results published by several authors[6,8] that boron tolerant genotypes accumulate less boron in the presence of both normal and high concentrations of boron and vice versa, while in our study tolerant genotypes accumulated less boron only when it was present in excessive concentration.

Table 3. Spearman's coefficients of rang correlation (r) among root length, dry weight and boron content on control and on boron treatments (average) in seedlings of 60 wheat genotypes

Pairs of traits	r
Control	
RL/W	0.61**
RL/BC	0.01 ns
W/BC	0.01 ns
Treatments	
RRL/RW	0.49**
RRL/RBC	–0.33**
RW/RBC	–0.26*

RL, W, BC-root length, dry weight and boron content in wheat seedlings on control; RRL, RW, RBC-relative root length, dry weight and boron content in wheat seedlings on boron treatments with regard to control; ns, *, **-insignificant, significant at the 0.05 and 0.01 levels of probability, respectively.

All pairs of treatments (B50/B100, B50/B150, B100/B150) concerning all analysed traits related to tolerance to excess boron (relative root length, relative dry weight and relative boron content in seedlings) were positively correlated (Table 2). Therefore, any applied boron treatment may be informative enough for further research on boron tolerance in wheat.

Nevertheless, the study was undertaken in laboratory and there was no possibility to investigate interactions between levels of boron tolerance and various environmental factors (drought, high temperatures etc.) which normally occur in field conditions. Field trial that is currently conducted at Institute of Field and Vegetable Crops in Novi Sad will highlight the impact of numerous environmental factors on boron accumulation and make possible the selection of genotypes which will be included in breeding programs.

4. Conclusion

Relative root length of wheat seedlings was used as selection criterion for boron tolerance. Out of sixty genotypes, fourteen were boron sensitive, seventeen medium sensitive, nineteen medium tolerant and ten genotypes may be considered as boron tolerant. Genotypes with small root growth suppression on treatments, namely boron tolerant genotypes, generally had smaller dry weight reduction and lower boron content.

Any analysed boron treatment may be informative enough for further research on boron tolerance in wheat, due to positive correlations found among diverse applied boron concentration treatments for all studied traits.

Acknowledgement

Milka Brdar is a scholarship holder of Ministry of Science of Republic of Serbia.

References

1. Y. Chantachume, D. Smith, G.J. Hollamby, J.G. Paull and A.J. Rathjen, *Plant Soil.* **177**, 249 (1995).
2. F.M. Eaton, *J. Agric. Res.* **69**, 237 (1944).
3. R.E. Ferreya, A.U. Aljaro, R. S. Ruiz, L.P. Rojas and J.D. Oster, *Agric. Water Manage.* **34**, 111 (1997).
4. C. Huang and R.D. Graham, *Plant Soil.* **126**, 295 (1990).
5. S.P. Jefferies, M.A. Pallotta, J.G. Paull, A. Karakousis, J.M. Kretschmer, S. Manning, A.K.M.R. Islam, P. Langridge and K.J. Chalmers, *Theor. Appl. Genet.* **101**, 767 (2000).
6. M. Kalayci, A. Alkan, I. Çakmak, O. Bayramoğlu, A. Yilmaz, M. Aydin, V. Ozbek, H. Ekiz and F. Ozberisoy, *Euphytica* **100**, 123 (1998).
7. G.D. Laing, F.M.G. Tack and M.G. Verloo, *Analytica Chimica. Acta.* **497**, 191 (2003).
8. R.O. Nable, *Plant Soil.* **112**, 45 (1988).
9. R.O. Nable, J.G. Paull and B. Cartwright, *Plant Soil.* **128**, 225 (1990).
10. R.O. Nable, G.S. Bañuelos and J.G. Paull, *Plant Soil.* **193**, 181 (1997).

11. J.G. Paull, B. Cartwright and A.J. Rathjen, *Euphytica* **39**, 137 (1988).
12. J.G. Paull, Genetic studies on the tolerance of wheat to high concentrations of boron, Ph.D. Thesis, Adelaide University (1990).
13. R.J. Reid, J.E. Hayes, A. Post, J.C.R. Stangoulis and R.D. Graham, *Plant Cell. Environ.* **27**, 1405 (2004).
14. C. Su, L.J. Evans, T.E. Bates and G.A. Spiers, *Soil. Sci. Soc. Am. J.* **58**, 1445 (1994).
15. *Statistica 7.0*, StatSoft, University License, University of Novi Sad (2004).

CHAPTER 17

THE EFFECT OF EXTRUDING ON *AMARANTHUS SP.* GRAIN AND GRITS QUALITY

MARIJA BODROŽA-SOLAROV, IRENA DOŠENOVIĆ AND BOJANA FILIPČEV

Institute for Food Technology, University of Novi Sad, Novi Sad, Serbia

Thermal treatment of cereals is used for improvement of their nutritive, hygienic, physic-chemical and other characteristics. *Amaranthus sp.* grains, which are similar in nutritive value and food processing to that of the cereals, have been classified to pseudocereals. The extruded products were obtained by extrusion of whole grains and grits of *Amaranthus sp.* The experiments were carried out on a single-screw extruder at a medium temperature (110°C) and humidity of 18%. Grains, grits and extruded products were analysed for chemical composition, energy value and microbiological status. The extrusion of *Amaranthus sp.* grain and grits decreased moisture and starch content, while the protein content remained constant. Thermal treatment improved the microbiological status of products in relation to untreated ones.

1. Introduction

Amaranthus sp. is one among a wide range of under-exploited crops showing promising potentials as a global resource of nutritious grains. The nutritional or health promoting ingredients in amaranth grain include lysine, an amino acid which is essential to body functions, squalene that has cholesterol-lowering properties, dietary fibre, tocotrienols and isoprenoid compounds. It has been reported to have three times higher fibre content than that of wheat, higher levels of calcium and iron as well as many other vitamins and minerals.[1]

Amaranth grain has been researched in several areas including — agricultural production,[2] potential uses, protein content[3] and evaluation of the whole grain.[4] The extrusion process is accompanied by changes of carbohydrate complex, namely by reduction of molecular weight of starch affected by degradation of starch into dextrin.[5] The small size of seeds and their oil content of approximately 8% allow extrusion without grinding or adding of additional water. By selecting the proper screw configuration along with a range of barrel temperatures and screw speeds, the extruded products with varying degrees of expansion can be obtained.

The work is aimed at examining the effect of extrusion on *Amaranthus sp.* whole grain and grits quality.

2. Material and Methods

Grains of *Amaranthus sp.* were obtained from local commercial producers. *Amaranthus sp.* grains and grits used in the investigation were extruded. Grits of *Amaranthus sp.* were obtained by milling on Bühler MLU 202 mill (granularity ranged from 650μm to 250μm). In this paper, we also investigated pellets produced by extrusion using single screw "E60" ("Sever" Subotica) extruder at 110°C and humidity of 18%.

The proximate chemical analyses of grits and extruded pellets were determined according to the valid regulations. The protein content (N × 5.8) was estimated by Kjeldahl method. Carbohydrate content was determined by the total starch content according to Ewers. The fat content was determined by Soxhlet extraction. The cellulose content was determined according to Wender. The moisture and ash content were determined by standard methods in accordance with the valid regulations.

Microbiological investigations included the following analyses: determination of the total aerobic mesophilic bacteria count per 1 g, total number of yeasts per 1 g, total number of moulds per 1 g, presence of *Salmonella spp*, Coagulase-positive staphylococci, Sulphite-reducing clostridia, *Proteus spp.* and *Escherichia coli.* Microbiological analyses were carried out according to the applicable regulations.

3. Results and Discussion

The extrusion is a thermal process for cereals treatment aimed at improving nutritive, sanitary and physicochemical characteristic of raw materials. The chemical composition of amaranth grains, before and after extrusion is presented in Figure 1. It significantly differed in moisture and starch content. The crude protein, crude fibre and ash content of *Amaranthus sp.* grain, before and after extrusion remained constant.

Chemical composition of *Amaranthus sp.* grits, before and after extrusion is presented in Figure 2. The chemical composition of *Amaranthus sp.* grits before and after extrusion did not significantly differ except for the moisture and starch content. The starch content decreased after extrusion probably due to the partial degradation of starch components, amylose and amylopectin. The obtained results are in accordance with the investigations carried out by Filipović *et al.* (2003).

The microbiological profile of amaranth grains, grits and extrudates are presented in Table 1. The application of thermal treatment contributed to the improved level of product sanitation.

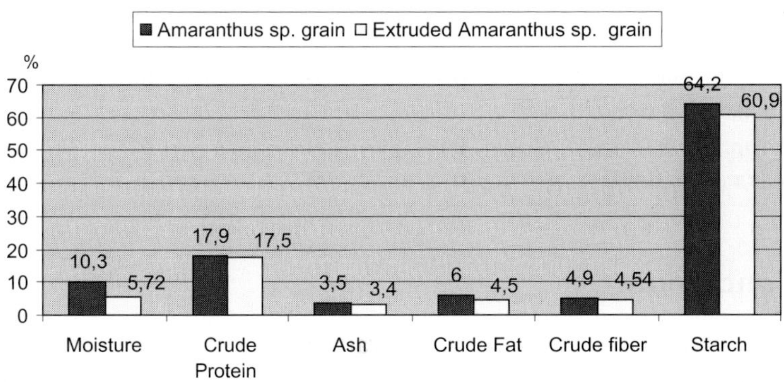

Figure 1. Chemical composition of *Amaranthus sp. grain* and extruded *Amaranthus sp. grain.*

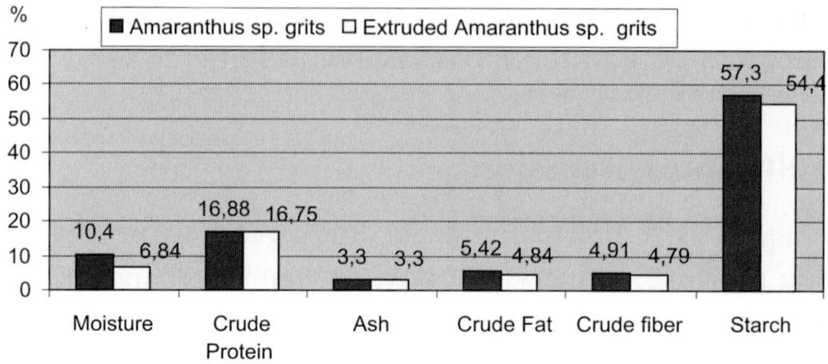

Figure 2. Chemical composition of *Amaranthus sp. grits* and extruded *Amaranthus sp. grits.*

Table 1. Microbiological status of *Amaranthus sp.* grains, grits and extrudates

Quality parameters	***Amaranthus sp. grits***	***Amaranthus sp. grain***	**Extruded *Amaranthus sp. grits***	**Extruded *Amaranthus sp. grain***
Total number of aerobic mesophilic bacteria per 1 g	10000	950	6000	200
Total yeast count per 1 g	0	0	0	0
Total mould count per 1 g	30	20	0	0
Salmonella spp. per 25 g	0	0	0	0
Coagulase-positive staphylococci per 0.01 g	0	0	0	0
Sulphite-reducing clostridia per 0.01 g	0	0	0	0
Proteus spp. per 0.001 g	0	0	0	0
Escherichia coli u 0.001 g	0	0	0	0

4. Conclusion

The extrusion of *Amaranthus sp.* grain and grits decreased moisture and starch content, while protein content remained constant. Thermal treatment improved the microbiological status of the product when compared to that of non-treated ones.

References

1. R.M. Saunders and R. Becker, *Adv. Cereal Sci.* **357**, (1984).
2. M. Bodroža-Solarov, Effect of genotype and sowing date on yield and yield components of the genus *Amaranthus* L., Ph.D. Thesis, Faculty of Agriculture, University of Novi Sad (2001).
3. R.D. Bressani, E.C.M. Martell and C.M. De Godinez, *Plant Food Hum. Nutr.* **43**, 123 (1993).
4. A. Sanchez-Marroquan, F.R. Del Valle, M. Escobedo, R. Avita, S. Maya and M. Vega, *J. Food Sci.* **51**, 1231 (1986).
5. R.M. Van Den Einde, C. Akkermans, A.J. Van Der Goot and R.M. Boom, *Carbohyd. Polym.* **56**, 415 (2004).

CHAPTER 18

EVALUATION OF ANTIOXIDANT ACTIVITY OF MEDICAL PLANT EXTRACTS AND THEIR APPLICATION IN BISCUITS

ALEKSANDRA MIŠAN, OLIVERA ŠIMURINA, ĐORĐE PSODOROV, MARIJANA SAKAČ, IVANA SEDEJ, ANAMARIJA MANDIĆ AND ŽARKO KEVREŠAN

Institute for Food Technology, University of Novi Sad, Novi Sad, Serbia

Medical plants have received much attention as a source of biologically active substances including antioxidants, antimutagens and anticarcinogens. In this paper, antioxidant activities of two mixtures of medical herbs, "Vitalplant" and "Gastroherb", products of the Institute of Medical Plant Research "Dr Josif Pančič" from Belgrade, were tested. Also, the mixtures were applied in wire-cut cookies and the antioxidant activity of products was tested. The added amount of herbal mixtures does not adversely affect the sensory quality of the product but provides specific aroma accepted by sensory panelists. The investigated mixtures as well as products exhibited an antioxidant activity.

1. Introduction

Current researches into free radicals have confirmed that foods rich in antioxidants play an essential role in the prevention of cardiovascular diseases, cancers and neurodegenerative diseases, as well as inflammation and problems caused by cell and aging.[1]

The onset of rancidity in baked products follows quite a different pattern from that of fat alone. Keeping of quality of crackers and cookies is of great economic importance since these products are often stored for extended periods before they are consumed, and they are not protected from oxidation.[2]

Spices and herbs have been added to foods since ancient times, not only as flavouring agents, but also as folk medicine and food preservatives. Spices and herbs and their constituents are generally recognized to be safe, either because of their traditional use without any documented detrimental impact or because of dedicated toxicological studies.[3] Being natural foodstuffs, spices and herbs appeal to many consumers who question the safety of synthetic food additives.

The objective of this study was to evaluate the antioxidant activities of ethanolic extracts of two mixtures of medical herbs, "Vitalplant" and "Gastroherb", which are composed of medical herbs that are known to have digestive stimulation action, as well as of cookies prepared, with addition of these herbal mixtures.

2. Materials and Methods

2.1. *Plant Materials*

Two mixtures of "Vitalplant" medical herbs (Rhamnus Frangula bark (35%), Mentha × piperita L. leaves (20%), Carum Carvi L. seeds (20%), Petroselinum crispum seeds (20%)) and "Gastroherb" (Origanum Vulgare L. (55%), Coriandrum sativum L. seed (30%), Cynara scolymus leaves (15%)), are the products of the Institute of Medical Plant Research "Dr Josif Pančič" from Belgrade. Herbal mixtures were in the form of powder with granulation of up to 3 mm.

2.2. *Preparation of Plant Extracts*

About 10 g of each plant material was shaken with 100 ml of 96% ethanol, at the room temperature for 1 h, and the mixture was left to stand overnight. The residues obtained after filtration were re-extracted twice with 96% ethanol by shaking for 1 h and filtrated. The combined extracts were evaporated to dryness in the rotary evaporator at 40°C, and stored at –4°C until further use.

2.3. *Wire-Cut Cookie Formulations*

The basic cookie dough formulation contained 100 g of white flour, 30 g of sucrose, 50 g of hydrogenated vegetable oil and required amount of water (~ 4 g). The dough was sheeted to a thickness of 3.5 mm and cut into circular shapes using 45 mm cutter and placed on an aluminium tray, baked at 180°C for 10 min and then allowed to cool. The cookies were stored at room temperature in polypropylene bags.

Cookies were prepared to provide three variations. The control treatment was prepared without plant mixture addition while the other two, cookies with herb mixtures, were prepared by adding 2 g of plant mixture to 100 g of dough. Initially, cookies were prepared with addition of different levels of herbs (0.5 g, 1 g, 2 g and 3 g to 100 g of dough), and were subjected to sensory acceptability. The most acceptable (2 g/100 of dough) was chosen for the experiment.

2.4. *DPPH˙ Free Radical-Scavenging Activity Test*

Plant extracts as well as cookies were tested in DPPH˙ free radical-scavenging activity test. Plant extracts were dissolved in methanol, in appropriate manner to obtain a series of dilutions (0.07–0.8 mg/ml).

Ground cookies (~ 2.5, 5, 7.5, or 10 g) and methanol (40 ml) were placed in the flasks, shaken at room temperature for 1 h, and the mixture was left to stand overnight. After filtration (Whatman No. 1 filter paper), the volume of the filtrate was adjusted to 50 ml with methanol. Methanolic extracts represented ~ 50, 100, 150 or 200 mg of cookie sample per millilitre.

Radical-scavenging activity against the stable radical DPPH˙ (1,1-diphenyl-2-picryl-hydrazyl radical (Sigma, USA) was determined spectrophotometrically following the procedure of Espin *et al.*[4] The IC_{50} (mg/ml) was defined as the concentration of an antioxidant extract which was required to quench 50% of the initial DPPH˙ under the given experimental conditions; it was obtained by interpolation from linear regression analysis.

2.5. *Antioxidant Activity (AOA) of Plant Extracts*

Antioxidant activity of plant extracts, based on coupled oxidation of ß-carotene and linoleic acid was determined according to the method of Moure *et al.*[5] Degradation rate of the extracts was calculated according to first order kinetics, and the antioxidant activity (AOA) was expressed as % inhibition relative to the control.[6] The IC_{50} (mg sample/ml) was defined as the concentration of an antioxidant extract which was required to inhibit the degradation of ß-carotene 50% relative to the control under the given experimental conditions; it was obtained by interpolation from linear regression analysis.

2.6. *Total Phenolics and Total Flavonoid Content of Plant Extracts*

The content of total phenolics in plant extracts, measured as gallic acid equivalents was determined using Folin-Ciocalteau´s reagent.[7] The flavonoids content in the extracts was measured by the $AlCl_3$ method based on the formation of a complex flavonoid-aluminium. Rutin was used for calibration curve preparation.[8]

2.7. *Reducing Power of Plant Extracts*

The reducing power was determined by measuring the formation of Perl's Prussian blue at 700 nm.[9] IC_{50} value (mg extract/ml) is the effective concentration at which the absorbance was 0.5 for reducing power and was obtained by interpolation from linear regression analysis. BHT (butylated hydroxy toluene) was used for comparison.

2.8. *Estimation of Lipid Oxidation in Cookies*

The extent of lipid oxidation was measured by thiobarbituric acid-reactive — substances assay (TBARS), based on reaction between malonyldialdehide MDA and thiobarbituric acid (TBA).[10]

2.9. *Statistical Analysis*

Experimental results are given as mean ± SD of three parallel trials and measurements.

3. Results and Discussion

The obtained results for total phenolics content, total flavonoids content, DPPH˙ scavenging activity test, AOA and reducing power of plant extracts are presented in Table 1.

The yields of ethanolic extracts of "Vitalplant" and "Gastroherb" mixtures were 10.07 and 10.61%, respectively. The total phenolic content as well as the total flavonoids content of each extract was estimated, since phenolics and flavonoids may significantly contribute to its overall antioxidant activity. The amount of total phenolics as well as the composition of the extract is highly dependent on extraction method and the type and polarity of extraction solvent.[11] Higher contents of total phenolics and flavonoids are found in "Gastroherb" plant mixture.

DPPH˙ free radical-scavenging activity of investigated plant extracts was highly concentration dependent. Both extracts showed linear regression ("Vitalplant" $y = 112.06 \times -0.7823$, $R^2 = 0.9946$, "Gastroherb" $y = 48.465 \times -5.1493$, $R^2 = 0.9932$) within the investigated concentration range (0.07–0.8 mg/ml). The obtained IC_{50} values indicate that ethanolic extract of "Vitalplant" plant mixture possessed better free radical-scavenging activity (Table 1).

Table 1. Total phenolics content, total flavonoids content, DPPH˙ scavenging activity test, AOA and reducing power of plant extracts

	Total phenolics content (gallic acid eq, % dry extract)	Total flavonoids (rutin eq, % dry extract)	DPPH˙ IC_{50} (mg/ml)	AOA IC_{50} (mg/ml)	Reducing power IC_{50} (mg/ml)
"Vitalplant"	0.696 ± 0.036	0.691 ± 0.095	0.453	2.26	2.56
"Gastroherb"	1.190 ± 0.011	0.898 ± 0.075	1.14	1.74	1.50

The reducing power is often used as an indicator of electron-donating activity, which is an important mechanism of phenolic antioxidant mechanism.[12] For the concentration range between 0.4–2 mg/ml, dependence between reducing power and concentration was linear ("Vitalplant" $y = 0.1714 \times +0.0618$, $R^2 = 0.9985$, "Gastroherb" $y = 0.2884 \times +0.0673$, $R^2 = 0.9775$). Obtained IC_{50} indicate that ethanolic extract of "Gastroherb" plant mixture possessed better reducing power (Table 1). IC_{50} for BHT was 0.24 mg/ml. "Gastroherb" plant mixture exhibited better AOA activity, which is indicated by lower IC_{50} (Table 1). Like for other antioxidant parameters good linearity between concentrations and AOA was achieved ("Vitalplant" $y = 0.045 \times +0.0145$, $R^2 = 0.9954$, "Gastroherb" $y = 0.0488 \times -0.7029$, $R^2 = 0.9883$), for the concentration range 0.17–2 mg/ml.

The IC_{50} values for $DPPH^{\cdot}$ free radical-scavenging activities of cookies are presented in Table 2. Within investigated concentration range good linearity was achieved in all investigated cases (control $y = 183.22 \times +19.756$, $R^2 = 0.9518$, cookies-"Vitalplant" $y = 292.35 \times +27.562$, $R^2 = 0.9406$, cookies-"Gastroherb" $y = 444.07 \times +31.232$, $R^2 = 0.9623$). As it can be seen in Table 2, the control possess a rather strong $DPPH^{\cdot}$ radical scavenging activity. That could be attributed to the Maillard reaction products.[13]

Despite the fact that "Vitalplant" ethanolic extract possessed better $DPPH^{\cdot}$ scavenging activity (Table 1), cookies supplemented with "Gastroherb" herbal mixture were better in the same test (Table 2). This might be due to synergism between antioxidants, or due to differences related to matrix.

The extent of lipid oxidation in TBARS test was performed after 6 weeks of storage at room temperature. MDA values (Table 2)

Table 2. $DPPH^{\cdot}$ scavenging activity test and MDA content in cookies

	$DPPH^{\cdot}$ IC_{50} (g/ml)	**MDA (mg/kg)**
Control	0.165	3.01 ± 0.079
Cookies-"Vitalplant"	0.077	2.77 ± 0.009
Cookies-"Gastroherb"	0.042	2.55 ± 0.002

indicate that both plant mixtures were capable of retarding lipid oxidation and that "Gastroherb" mixture was more efficient.

4. Conclusions

Ethanolic extracts of "Vitalplant" and "Gastroherb" plant mixtures exhibited antioxidant activities in all investigated assays. Plant mixtures added to cookies were capable of retarding lipid oxidation of cookies. Cookies supplemented with plant mixtures possess higher ability to scavenge free radicals than the control.

Regarding all the results, it can be concluded that "Vitalplant" and "Gastroherb" plant mixtures as well as cookies prepared with these herbal mixtures could be regarded as a potential functional food.

References

1. T. Finkel and N.J. Holbrook, *Nature.* **408**, 239 (2000).
2. S.S. Bassiouny and F.R. Hassanien, *Food Chem.* **37**, 297 (1999).
3. E.J. Smid and L.G.M. Gorris, Natural antimicrobials for food preservation, *Handbook of food preservation*, Marcel Dekker, 285 (1999).
4. J.C. Espin, C. Soler-Rivas and H.J. Wichers, *J. Agric. Food. Chem.* **48**, 648 (2000).
5. A. Moure, J.M. Cruz, D. Franco, J.M. Domínguez, J. Sineiro, H. Domínguez, M.J. Núñez and J.C. Parajó, *Food. Chem.* **72**, 145 (2001).
6. M.S. Al-Saikhan, L.R. Howard and J.C. Jr. Miller, *J Food. Sci.* **60**, 341 (1995).
7. V.L. Singleton, R. Orthofer and R.M. Lamuela-Raventos, *Method. Enzymol.* **299**, 152 (1999).
8. V. Dewanto, X.Wu, K.K. Adom and R.H. Liu, *J. Agric. Food. Chem.* **50**, 3010 (2002).
9. M. Oyaizu, *Jpn. J. Nutrit.* **44**, 307 (1986).
10. D.M. Hodges, J.M. DeLong, C.F. Forney, R.K. Prange, *Planta.* **207**, 604 (1999).
11. J.K.S. Moller, H.L. Madsen, T. Aaltonen and L.H. Skibsted, *Food. Chem.* **64,** 215 (1999).
12. A. Yildirim, A. Mavi and A. Kara, *J. Agric. Food. Chem.* **49**, 4083 (2001).
13. H. Jing and D.D. Kitts, *Food. Res. Int.* **33**, 509 (2000).

CHAPTER 19

FATTY ACID COMPOSITION OF TSIGAI SHEEP MILK AS A PHYSIOLOGICAL ADVANTAGE

JÓZSEF CSANÁDI, JÓZSEF FENYVESSY AND ILDIKÓ BAJÚSZ

Faculty of Food Engineering, University of Szeged, Szeged, Hungary

According to many authors sheep milk has nutritional advantage in relation to cow milk. Some authors explain this by the different fatty acid (FA) composition, mainly the higher ratio of unsaturated (UFA) and C4–12 fatty acids. Nowadays, however, it is very important to know the amount and the ratio of n-6 and n-3 fatty acid in our foods. Our objective was to investigate the fatty acid composition of milk from the milking Tsigai sheep. The obtained results show that physiological evaluation of FA composition of Tsigai sheep milk and its products is more favourable in every respect than cow milk.

1. Introduction

The beneficial effect of ruminant's milk and milk products for human health is well known. But, small differences in milk composition can cause noticeable variations in physiological effects. The difference in the fatty acid composition of sheep, goat and cow milk is a good example for it. According to some authors the fatty acid composition of the sheep milk is similar to cow milk.[10]

Other authors have found noticeable differences in fatty acid composition in relation to the cow milk.[4–7,9] According to some authors more favourable physiological determination of the sheep milk fat as compared to the cow milk can be explained by the higher ratio of unsaturated and C4:0 12:0 fatty acids.[2,3] Milk and majority of milk products have another special nutritional advantage because they contain the optimum ratio (3:1) of n-6 to n-3 fatty acids. Our objective was to investigate the fatty acid composition of milk from the milking Tsigai sheep.

2. Material and Methods

The fatty acid composition of milk was determined from bulk milk samples in the whole (165 day) lactation. The forage of the flock was based on grazing and was of characteristically extensive type. We analysed separated milk fat samples from daily milk samples (morning + evening milking) stored at 25°C until the analysis.

The determination of fatty acid composition:

1. Preparation: The samples were destroyed in a hot water bath with concentrated hydrochloric acid and mixed with ethanol. Afterwards, the lipids were extracted with ether and petrol ether (< 60°C). After combining the organic phases the solvent was removed by means of a rotating vacuum evaporator.
2. Hydrolysis and esterification: The evaporated samples were boiled with 0.5M methanol sodium hydroxide solution (app. 5 minutes) and the boiling was furthered continued for 3 minutes with 14% methanol boron-trifluoride solution. Boiling was then extended for 1 minute with the addition of dried hexane and after cooling down, mixed with salt water solution. After separation of phases, 0.5–2 μl samples were taken from the organic phase and injected into Chrompack CP 9000 gas chromatograph.

3. Results

The Tsigai's milk fat contains more saturated than unsaturated fatty acids, as shown in the reference literature, and the difference was 18.27%. Tsigai's milk fat also contains other PUFA, such as C20:2n 0.46%, C20:3n3 0.12%, C20:4n6 0.15% and C22:2n 0.15%.

The ratio of unsaturated fatty acids (UFA) to polyunsaturated fatty acids (PUFA) ratio in total fatty acids was 40.81% and 3.82%, respectively. The amount of C18:1n9c (29.98%) was the highest in all FA, being higher by about 4.0% than in cow milk fat. SFA/UFA ratio and the amount of PUFA confirm that the sheep milk has advantages for human nutrition, as shown in the Figure 1 below.

The change in the ratio of different FA groups was the smallest in PUFA groups (0.36%), while in other groups it was similar 11.37–11.52%, although the SD% of SFA was only 8.24%. The ratio of UFA was higher if SFA's value was lower in each case and the value of UFA was the highest in May, what was related to feeding.

The data showed in Table 2, confirmed that the fatty acid composition of Tsiagi's milk fat is very close to the Hypothetical Ideal Fatty Acid composition. The ratio of SFA, UFA and oleic acid meets the requirements of HIF and the ratio of short chain fatty acids is very close to it. It is to be noted that the milk fat contains essential fatty

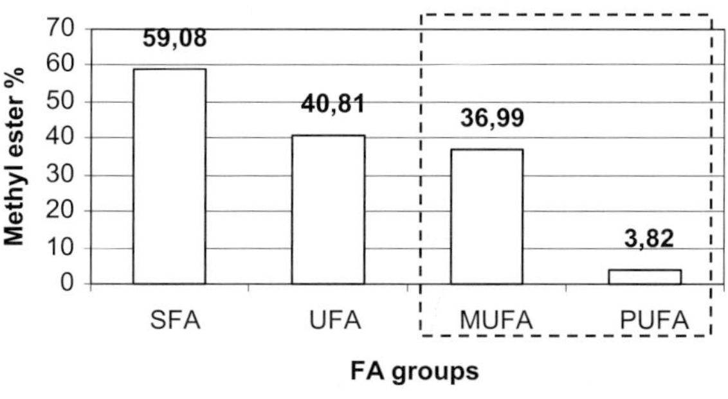

Figure 1. Different nutritional fatty acid groups in Tsigai sheep milk fat.

Table 1. The nutritional fatty acid groups in the lactation (n = 6)

FA groups	Mean	SD	SD%	Max	Min
UFA	40.81	4.91	12.03	45.26	33.74
SFA	59.18	4.88	8.24	66.16	54.69
MUFA	36.99	4.81	13.01	41.34	29.97
PUFA	3.82	0.14	3.55	4.00	3.66

Table 2. Comparison of the Tsigai sheep milk fat to Hypothetical Ideal Fatty acid composition (HIF)

Fatty acids	HIF %	Tsigai %
Saturated fatty acids	53.0–62.0	59.08
Short and medium chain fatty acids in total saturated FA (C4–C12)*	10.0–12.0	9.23
Unsaturated fatty acids	38.0–47.0	40.81
Oleic acid (C18:1n9c)	28.0–32.0	29.98
Linoleic acid (C18:2n6c)	7.0–12.0	2.15
Linolenic acid (C18:3n3)	0.5–1.0	0.75

acids, such as linoleic and linolenic acids, but the ratio of these fatty acids, particularly the linoleic acid, does not reach the requirements, whereas the latest researches have reported that the optimum (n-6)/(n-3) ratio is lower than one showed in the Table 2. Therefore, Tsigai's milk fat meets the requirements of HIF (as the milk fat from different species) except that it contains about 5.0% less essential fatty acids, particularly linoleic acid.

Today, the usual diet in industrial countries contains much less ω-3 fatty acids than the diet of a century ago. The diet from a century ago had much less ω-3 than the diet of early hunter-gatherers.[8] We can also look at the ratio of ω-3 to ω-6 in comparisons to their diets. These changes have been accompanied by the increased rates of many diseases — the so-called diseases of civilisation — that involve inflammatory processes. There is now very strong evidence[5] that several of these diseases are ameliorated by increasing dietary ω-3, and good evidence for many others. There is also more preliminary evidence showing that dietary ω-3 can ease the symptoms in several psychiatric disorders.[1] For these reasons it is very important to investigate the amount and the ratio of n-6 and n-3 fatty acids in foodstuffs and raw materials.

Tsigai's milk fat contains n-6 and n-3 fatty acids as shown in Table 2. Regardless the fact that the (n-6)/(n-3) ratio in Tsigai's misfit is the optimum from the human nutrition standpoint, the consumption of milk products *per se* cannot compensate the

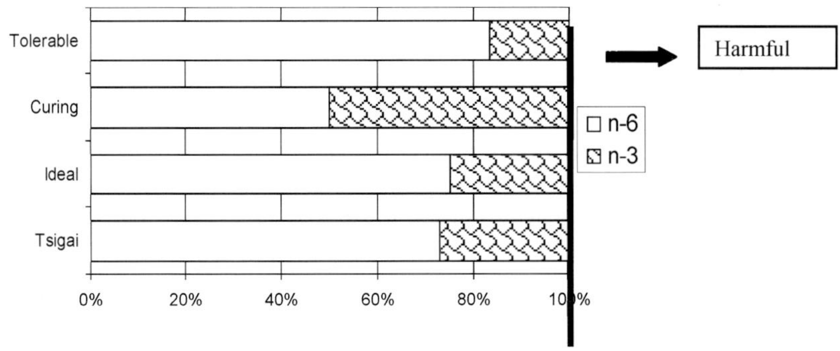

Figure 2. The place of Tsigai sheep milk fat related the ω-6/ω-n3 fatty acid ratio in the diet.

predominance of n-6 fatty acids in traditional nutrition in most of the European countries. From the nutritional viewpoint, the presence of about 80% of α-linolenic acid in Tsigai's milk fat has the most beneficial effects.

4. Conclusion

The fatty acid composition of Tsigai milk does not significantly differ from other data published in specialised literature (mainly in the last decade), but it has some beneficial advantages. Tsigai milk fat is close to the hypothetically ideal fatty acid composition. Moreover, it has the optimum ω-6/ω3 fatty acid ratio (2.68). The SFA/UFA ratio in Tsigai sheep milk fat is the optimum as in the goat and cow milk, but the amount of total unsaturated fatty acid is somewhat higher.

The oleic acid ratio (C18:1n9c) is noticeably higher in Tsigai sheep milk than in milk from other sheep genotypes. The studied differences confirm that sheep milk (milk fat) has more beneficial advantages for human nutrition than cow milk and fats from other foods, in particular (e.g. meat or sunflower oil). The consumption of the sheep milk products need to be promoted in Hungary, too, to allow the consumers to enjoy the benefits of the FA composition of sheep milk.

References

1. R. De Caterina and G. Basta, *Eur. Heart. J. Suppl.* **3**, 498 (2006).
2. J. Fenyvessy and J. Csanádi, *J. Acad. Hungarian Dair.* **2**, 23 (1999).
3. G.F.W. Haenlein, *Int. J. Anim. Sci.* **16**, 253 (2001).
4. W.R. Morrison, *Lipids* **3**, 101 (1968).
5. Omega-3 fatty acids, fish oil, alpha-linolenic acid., National Institute of Health (2005).
6. L. Posati and M.I. Orr, Agriculture Handbook 8-1, Composition of Foods: Dairy and Egg Products, USDA, Washington (1976).
7. W.N. Sawaya and W.J. Safi, *Milchwissenschaft* **39**, 2 (1984).
8. W.N. Sawaya and J.K. Khallil, *Milchwissenchsaft* **40**, 81 (1985).
9. A. Simopoulos, *World Rev. Nutr. Diet.* **88**, 18 (2001).
10. Y.W. Park, M. Juárez and M. Ramos, *Small Ruminant. Res.* **68**, 88 (1979).

PART 3

THE ECOLOGICAL ISSUES

CHAPTER 20

ECOLOGICAL MODELLING: CHALLENGES AT A CROSSROADS

DRAGUTIN MIHAILOVIĆ

Faculty of Agriculture, University of Novi Sad, Novi Sad, Serbia

Ecological modelling has existed since the times when ecology was established as a scientific discipline. This approach in ecology has been developed from simple and rough techniques to very sophisticated modelling tools. In modelling procedures the ecologists are interested in the relationship between plants and the environment — the relationship that spans length scales from the molecular level (in the case of photosynthesis) to several kilometres and more (when considering energy flow in plants fields). In the situation when a man intrudes he inevitably disturbs the delicate balance of organisms, substances and their activity that nature has evolved. Ecosystems can adapt to a man's violent activities, but only up to a certain point. The impact on ecosystems can only be described quantitatively with the application of ecological models and systems ecology and their reactions to perturbations can only be understood via the ecosystem theory. Ecology has always been in an exclusive group of scientific disciplines that ask philosophy, biology, mathematics and physics to be supplied with their latest achievements. This comes from the fact that theoretical description of any ecological system has at least two crucial points: (a) to construct a concrete model of both the admissible states of the system and its transition between them and (b) to establish the rules of selecting among many theoretically admissible states of the system only those states that are realised in nature under the given external conditions. Therefore, ecology has always been at a crossroads having a destiny similar to those of the afore-mentioned disciplines. In this paper we will give a very short overview of the current state in ecological modelling from the complex systems point of view. First, we will make some comments about ecological complex systems underlying some of their main features. Second, we will point out the approaches used in modelling complex systems, particularly emphasising the category theory as the most general formalism in modelling of these systems. Finally, some concluding remarks will be drawn pointing out the current and further challenges in ecological modelling.

1. Introduction

In the modelling procedures the ecologists are interested in the relationship between plants and the environment — the relationship that spans length scales from the molecular level (in the case of photosynthesis) to several kilometres and more (when considering energy flow in the plants fields). Therefore, in modelling quantities of the plant environment we must bear in mind that: (i) spatial variability in the plant environment is highly related to variability in soil moisture, incoming solar radiation and stand structure; (ii) a non-linear response of many biological and physical processes in the plant environment to the modelled quantity means that simple calculation based on temporal averages can produce large errors; (iii) the knowledge of local environmental quantities variability in uniform plant stands is also required in sophisticated modelling of biophysical processes in the plant environment; and (iv) the afore-mentioned key ecological modelling points reflect additionally on the model reliability through computing procedures. Ecological modellers in numerically oriented studies base their calculations on mathematical models for simulation and prediction of different processes which are exclusively non-linear, describing relevant ecological quantities.[1] A theoretical description of any ecological (natural) system includes, at least, two important aspects. First, one should construct a concrete mathematical model of both the admissible states of the system and its transitions between these states. Second, one should establish the rules of selecting among many theoretically admissible states of the system

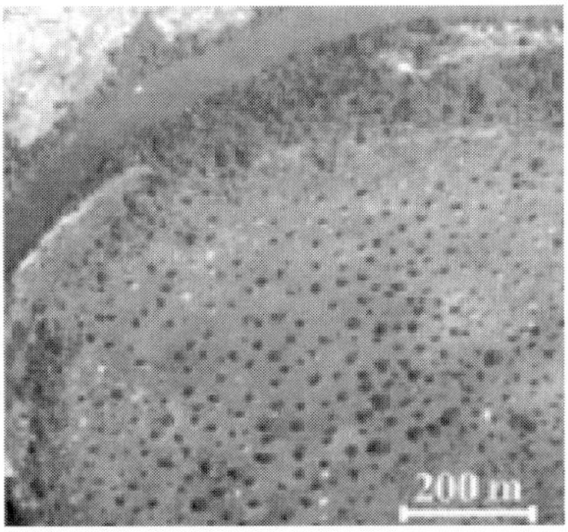

Figure 1. Pattern of isolated dense tree patches that marks the transition from tropical rain forest to grasslands in Marahoue´ National Park in Ivory Coast, West Africa: woody vegetation is dark grey, whilst grassland is light grey.[3]

only those states which are realised in nature under the given external conditions.[2] Ecological world is an ideal space for considering the complex systems. Within complex systems, each system, as an open one, interacts in a coherent way, producing new structures, building up cohesion and new structural boundaries see Figure 1. It goes through the emergence and self-organisation in the sense that (a) emergence means the appearance of a new property of a system which cannot be deduced or previously observed as a functional characteristic of the system and (2) self-organisation means the appearance of new system structures without the explicit pressure from outside the system, or involvement from the environment.

In this paper we will point out the general strategy in complex systems modelling, emphasising in particular the category theory as the most general formalism in modelling of these systems. Then we will draw some concluding remarks that point out the current and further challenges in ecological modelling.

2. Use of the Category Theory in Ecological Modeling

Figure 2, which was adapted from Rosen[1] (see Figure 2.3.1, p. 74), represents the modelling relation. It schematically depicts a modelling

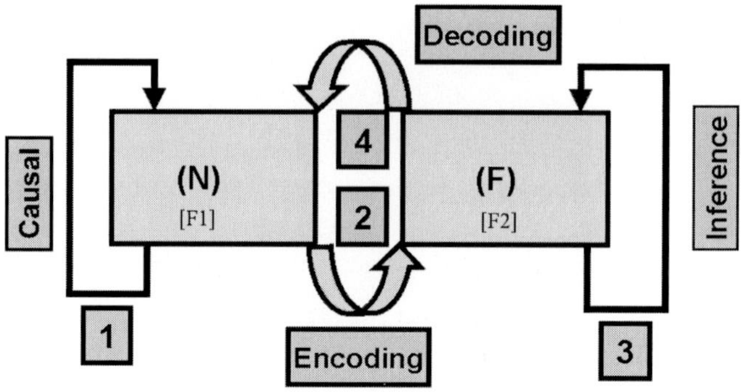

1- Causal entailment within natural system (N)
2- Encoding: observer's propositions about N are used as as hypotheses in construction of formal system (F)
3- Generation of theorems in F which function as a model of N
4- Decoding: application of theorems in F back to the N in the form of predictions

Figure 2. Schematic diagram representing both (i) the comparison of two formalisms F_1 and F_2 and (ii) modelling relation when we have given a natural system N, and a formal system F.[6]

relation when a natural system (N) and a formal system (F) are given. As above, two arrows represent the respective entailment structures; inference in formalism (F) and causality in a natural system (N). Now, the two established dictionaries provide encoding the phenomena of N into the propositions of F and another for decoding the propositions of F back to the phenomena in N. As mentioned above, there are two paths in diagram (1) and (2) + (3) + (4). According to the first of them (the path (1)) represents the causal entailment within N (what an observer will see by simply sitting and watching what is happening).[1] The arrow (2) encodes the phenomena in N into the propositions in F. In this route we must use these propositions as hypothesis based on which the inferential machinery of the formal system F may operate (denoted by the arrow (3)); it generates theorems in F, entailed precisely by the encoded hypotheses. Finally, we have to decode these theorems back into the phenomena of N, via the arrow (4). At this point, the theorems become *predictions* about N.

Then the formal system F is called a *model* of the natural system N if we always get the same answer regardless of the fact whether we follow path (1) or path (2) + (3) + (4). This Rosen's diagram is a recognisable guide for modelling community dealing with complex systems. The process of modelling complex systems is a very comprehensive one.

A system is to be treated as a complex structure, as for instance in Peter Checkland's definition: A system is a model of a whole entity; when applied to human, the model is characterised fundamentally in terms of hierarchical structure, emergent properties, communication and control (Ref. 4, p. 318). The major components of complexity are openness and freeness, but the distinctive characteristic is "natural activity" like self-organisation, and still of great importance as intra-activity but now joined by the phenomena of anticipation[5] and interactivity between systems to be found in global interoperability. The transition from connectivity to activity involves a type change and therefore requires formal system with an inbuilt facility to cross between the levels. Thus, intraconnectivity between the components cannot give rise to interactivity between those components without some non-local integrity coming into play.[7] The non-locality is a principle that is, among some others, specific for a certain object area such as the inorganic, living or human realm. This principle means certain interaction between the elements of the system that is treated as the transmission of information at infinite speed (see, for instance, Ref. 8). Category theory recommended by Robert Rosen as a modern tool for complex (and living) systems is found to have a formal expressive power as the process beyond modelling for exploring the fundamental non-local concept of adjointness needed to understand advanced systems.

The category theory (or topos) is recommended for modelling of biophysical systems and consequently the ecological systems. It provides both the mathematical modelling of physically observable structures, and logic conceptualisation of this modelling, at the same time. A category consists of: (1) a class of *objects*; (2) a class of morphism ("arrows") so that for each morphism, f, one object as the domain of f and one object as the co-domain off, see Figure 3(a); for each object, A, an identity morphism which has the domain A and co-domain A.

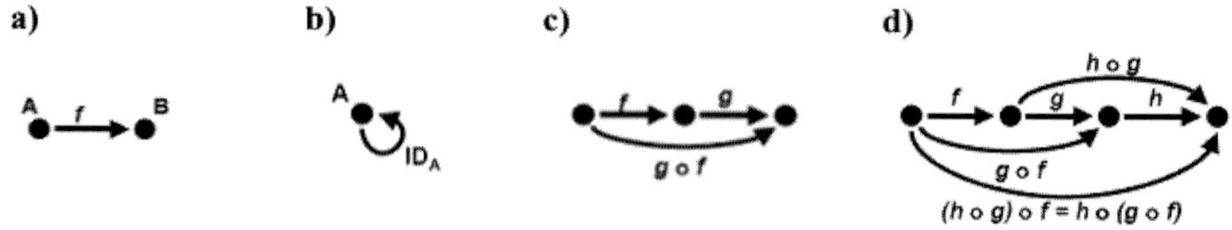

Figure 3. Schematic diagram representing the category theory essentials (a) morphism (arrows), objects, domain and co-domain, (b) identity morphism, (c) composite morphism and (d) identity composition and associativity (from Ref. 9).

("IDA") as seen in Figure 3(b) and (3) for each pair of morphisms f: A → B and g:B → C, (i.e. cod(f) = dom(g)), a composite morphism, g ° f: A → C , see Figure 3(c). With these rules we have: (1) for each morphism *f*: A → B, *f* ° ID_A = *f* and ID_B ° *f* = *f* (*identity composition*) and (2) for each set of morphisms f: A → B, g: B → C, h: C → D, (h ° g) ° f = h ° (g ° f) (*associativity*) as it is depicted in Fig. 3d. The arrow of category theory has not only formal meaning. According to Ref. 10 it formalises the principle of *constancy* (originally introduced by Heraclites and Parmenides) that is provided by a common source and target. Such an arrow is referred where source and target are indistinguishable. In a defined system, the collection of entities can be identified as objects while operations between them are defined by arrows. If we look at Figure 3(a) it is seen that there may be many possible arrows between objects. However, in category theory it is shown that a unique limiting arrow may exist for all these possible arrows that represent the resulting *intraconnectivity* of a local system. There is an ordering between the two entities given by the direction of arrows.[7] This means that the arrow limit between the two entities is also a limit of all possible paths. Because of the existence of limits and all possible connectivity, this is classified by axiomatic categories as a Cartesian closed category.

In category theory, it can be considered the category in which the objects are categories and the morphism are mappings between categories. The morphisms in such a category are known as *functors*. Given two categories, **C** and **D**, a functor **F**: **C** → **D** maps each morphism of **C** onto a morphism of **D**, so that, see Figure 4: (1) **F** preserves identities, i.e. if × is a **C**-identity, then

Figure 4. Schematic diagram representing the functor "action".

F(x) is a **D**-identity and (2) **F** preserves composition, i.e. **F**(f ° g) = **F**(f) ° **F**(g).

Moving up for one level, there is a grand limiting arrow for all the afore-mentioned limits existing as an identity functor characterising the type and therefore the system as a category. A system as a category may than be drawn as the circular arrow that is the identity functor identifying the type of a system.[7,10] The system is therefore one large arrow, i.e. the process in which the internal arrows are just components of one arrow. This then leads to *interconnectivity* between the systems. Also, the functor between two categories is conceptually the same as internal arrows between the arrows above. In the framework of this theory it is possible to repeat abstraction to one higher level (a natural transformation) but no higher. This level is the level of *interactivity.*

Let us note that self-organisation of a category theory-system (intra-activity) arises when the category-system pair is indistinguishable. Finally, the category theory is a very useful tool when we meet difficult problems in some areas of mathematics, ecology, physics, computer sciences, biological nano-engineering and self organisation of cell function in living systems,[11] among many others. They can be translated into (easier) problems in other areas (e.g. by using functors, which map from one category to another).

3. Instead of Conclusion

Ecological world is a space that is rich with complex systems. Within them, each system, as an open one, interacts in a coherent way, producing new structures, building up cohesion and new structural boundaries. It goes through emergence and self-organisation. Philosophy of modelling approaches in ecology is also rich but it now makes longer steps because of use of the category theory originally proposed by Robert Rosen to be used in modelling the complex ecological systems. Within the last fifty years, this theory, as the most general (non-axiomatic and self-closing type theory avoiding Russell's paradox) of modelling theories, made remarkable contributions at many fronts of ecology and related sciences — from micro to global scales. Its further efficient use in ecological models is closely connected with the development of mathematics, biology, physics, computer sciences as well as philosophy.

References

1. R. Rosen, Life itself, *A Comprehensive Inquiry into the Nature, Origin, and Fabrication of Life*, 285. (Columbia University Press, 1991).
2. A.P. Levich and A.V. Solov'yov, *Cybernet. Syst.* **30**, 571 (1999).
3. O. Lejeune and M. Tlidi, *Phys. Rev. E.* **66**, 010901.1 (2002).
4. P. Checkland, *Soft Systems Methodology, A 30-year Retrospective, Thinking Systems Practice* (John Wiley, 1981).
5. G.J. Klir, *AIP Conf. Proc.* **627**, 37 (2002).
6. D.T. Mihailovic and I. Balaz, *Idojaras* (2007).
7. N. Rossiter and M. Heather, Conditions for interoperability, *7th International Conference of Enterprise Information Systems (ICEIS)*, Florida, USA, 92 (2005).
8. J.S. Bell, *Physics* **1**, 195 (1964).
9. S. Easterbrook, http://www.cs.toronto.edu/~sme/presentations/cat101.pdf (1999).
10. E.G. Manes and M.A. Arbib, Arrows, structures and functors, The categorical imperative (1975).
11. O. Wolkenhauer and J. Hofmeyr, *J. Theore. Biol.*, 10.1016 (2007).

CHAPTER 21

USE OF FORMAL CONCEPT ANALYSIS FOR CONSTRUCTION OF SUBJECTIVE INTERFACE BETWEEN BIOLOGICAL SYSTEMS AND THEIR ENVIRONMENT

SINIŠA CRVENKOVIĆ

Faculty of Sciences, University of Novi Sad, Novi Sad, Serbia

DRAGUTIN T. MIHAILOVIĆ

Faculty of Agriculture, University of Novi Sad, Novi Sad, Serbia

IGOR BALAŽ

Institute of Physics, Belgrade, Serbia

Modelling of complex natural systems (either physical, chemical or biological) usually deals with estimation of dynamics of chosen processes and consequent analysis of the obtained data. In such an approach, one of the main premises is the supposed objectivity of the universe. For purely physical systems this can be hold as true, but in a realm of biological systems it can be treated only as a very crude approximation valid for a limited number of modelling purposes. Because of that, in order to establish more accurate estimation of pattern of interactions of biological systems with their environment, as well as interactions among living systems, it is necessary to take into account the manner in which they "observe" the environment, separate it into different functional patches and associate them with internal functional schematism. Formal representation of such processes can be elegantly and efficiently done using Formal Concept Analysis which is a branch of the applied lattice theory. Within that formal framework, the universe is separated into two sets, G (objects) and M (attributes), which can form a formal context (G, M, I) where I is a binary relation between G and M. In biological (non-cognitive) systems, attributes used for separating environmental factors one from another, are scarce in number and they can be divided into two main categories: physical influences (various types of radiation or temperature) and chemical influences (various types of molecules or ions). However, these attributes are further divided into a set of sub-attributes which can be represented as a scale of values within a given attribute. In that way, a segment of the environment is encircled by a certain attribute and is further divided according to the given scale within that attribute, thus establishing a many- valued context (G, M, W, I) where G is a set of objects, M is a set of many-valued attributes, W is a set of attribute values and I is a ternary relation $I \subseteq G \times M \times W$. Giving a simple abstract example, in this paper we will illustrate the basic principles of such process. The aim of this paper is to clearly show that even using a limited set of attributes (observational filters) it is possible, for a biological system, to efficiently reduce the enormous complicity of its objective environment into a functionally meaningful subjective construction.

1. Introduction

In the last few decades, mathematical modelling of biological systems became a very fast growing field. In order to understand a vast amount and complexity of data obtained by various biological or biochemical experiments it is necessary to try to formalise them and construct mathematical models which can help us to predict the outcomes of some processes or elucidate underlying mechanisms of some observed phenomena. Especially interesting field for a great number of scientists is biomedical modelling, which can range from various cellular processes (e.g. cancer growth, angiogenesis, tissue reparation, etc.) to the dynamics of interacting populations in epidemiology. However, almost all of these efforts are divided into two main branches: (i) modelling of particular, isolated processes within an organism and (ii) population modelling. Both of these approaches share the same premise — supposed objectivity of the observed data. When dealing with physical or even chemical systems this can be held as true, but in a realm of biological systems it can be treated only as a very crude approximation valid for a limited number of modelling purposes. If we determine our task, within strictly pragmatic boundaries, as a prediction yielding approximation of some particular cellular processes or outcomes of some population dynamics, there will be no problems in applying currently used modelling tools. But, on the other hand, if we define our task in a much broader terms, i.e. if we define it as a modelling of living systems in general, we should inevitably postulate completely different premises as starting points in our journey. In that case, our primary goal is rather not to predict some existing cytosolic or morphogenic process, but to catch pattern of dynamics and organisation of a cell (and organism) as a whole.

One of the main premises which should be changed in that case is the supposed objectivity. Instead of that, we need to introduce the notion of an observer which resides within the universe. This notion is not completely new in science, and an umbrella-term has been defined as endophysics,[1] but still requiring attention devoted to its stronger consistency. However, we can clearly define some important qualities of such observer-based systems.

First, their environment is defined from the perspective of that system. In other words, if we know that observation is not a simple act but a process of assimilation of external changes with internal operative schemes, then, in modelling, we can make a clear distinction between the construction of the environment (by observer) and the condition of that environment. An issue here is only a construction, i.e. modalities of perceiving the environment by living systems.

Second, it is a fundamental fact that each perceptive entity can observe or register only those changes in its vicinity which activate its functional operations. In that manner, the externality of any such system is pre-structured in accordance with its functional purposes and objectives. The mere fact that something can be perceived means that it becomes connected with the needs and functions of the living system. Conversely, since each organism constructs its environment as a field of interests, only functionally relevant changes can appear in it.

As an inherent consequence of such assimilation of reality, the external medium of living systems is transformed into a functionally treatable environment where external, arbitrary events can be conveyed into the internal scheme of reactions.

Third, such structuring (the interrelation of parts in an organised whole) of the environment is not directed only from organism to externality, but it also takes place within living systems where each specific form of interactions constructs its own space. Regarding organised systems in general, Luhmann[2] defined that problem as a need for replacement of classical paradigm parts-whole with paradigm of layered structuration which could be defined as a system-environment distinction. Applying this new paradigm does not by itself imply any kind of given hierarchies or primacies within a system, but it should be brought into the focus as a problem of local generation of perspectives where each subsystem constructs its own environment and its own system-environment distinction.

Because of all these facts, in order to establish more accurate estimation of pattern of interactions of biological systems with their environment, as well as interactions among the living systems, it is necessary to take into account the manner in which they "observe" the environment, separate it into different functional patches and associate them with internal functional schematism. Formal representation of such processes can be elegantly and efficiently done by using the Formal Concept Analysis (FCA), which is a branch of the applied lattice theory. FCA has been introduced by Wille[3] and it defines a concept as a unit of two parts: extension and intension. The extension covers all objects belonging to a particular concept and the intension comprises all attributes valid for all those objects. Both attributes and objects are united by a triple (G, M, I), which is called a formal context if G and M are sets and $I \subseteq G \times M$ is a binary relation between G and M.[4] Also, between these two closure systems (G and M) a dual isomorphism is established. The ordered set of all formal concepts of (G, M, I) forms the concept lattice of (G, M, I) which are always complete.

In this paper we will give a simplified but illustrative example of an abstract living system's interaction with the environment.

Our goal is to represent how subjective processing of the environment is achieved, what are the functional consequences and how this process influences dynamics of the organisation of living systems.

2. Construction of the Model

Let us define set $O = \{o_1, o_2, o_3 \ldots o_n\}$ as a set of all environmental objects. In order to functionally treat them, living systems need to associate each external object with some attribute. Hence, we define set of attributes $M = \{m_1, m_2, m_3 \ldots m_n\}$. Within FCA, objects and attributes are mutually defined, so we automatically obtain set $G \subseteq O$: $\{\forall g \in G | (g, m) \in I\}$, where I is a binary relation between M and G. In other words, every object (attribute) which does not form a binary relation with corresponding attribute (object) simply does not exist from the perspective of particular formal context. In that way, the concept of the "whole" environment changes its epistemological status and is reduced to an observable environment, i.e. to a set of treatable objects. The next step in functional configuration of the environment is separation of homogenous observable environment into a set of concepts within a given context defined by (G, I, M). In biological (non-cognitive) systems, the attributes used in separating environmental factors from one another, are scarce in number, and can be divided into two main categories: physical influences (various types of radiation or temperature) and chemical influences (various types of molecules or ions). However, these attributes are further divided into a set of sub-attributes which can be represented as a scale of values within a given attribute. In that way, a segment of the environment is encircled by a certain attribute and is further divided according to the given scale within that attribute, thus establishing many-valued context (G, M, W, I) where G is a set of objects, M is a set of many-valued attributes, W is a set of attribute values and I is a ternary relation $I \subseteq G \times M \times W$.

Here, we will use a very simple model of typical photosynthetic organism. Like in all living cells, an interaction with the environment is performed through currently active receptors, mainly located in a cell membrane. According to their configuration they can be divided into photo-receptors and receptors for various types of external molecules. It is important to emphasise that the activity of molecular receptors is always connected with successive configuration changes of these receptors and, the so-called, active transport through membrane. On the other hand, water, along with the dissolved ions, enters the cells spontaneously, either by diffusion, or through aquaporins. Therefore, in our conceptual scheme we postulate the existence of two main attributes: radiation and molecule (Table 1). The attribute "molecule" is further divided into two sub-classes according to the size of objects to which it can be applied. In the group of objects smaller than the supposed threshold there is water and small ions dissolved in it, while all molecules larger than the threshold are in the second group (denoted by Larger than[(1)] and Smaller than[(2)], respectively). Further diversification of attributes is done by defining several types of radiation stimuli and several types of recognisable molecular structures. Finally, we introduced one more attribute, defined as a global regulator, which, in order to preserve simplicity of the model, does not have any sub-attributes. Global regulators are ubiquitous in living systems and they are activated in response to almost all kinds of external stimuli. Although objects in this model correspond to physical objects, no specific description is assigned to them because they gain full "subjective" meaning only through construction of

Table 1. List of all the attributes used in the model. In further text, the term Structure (X … Q) will be denoted by the strings Str(X … Q)

Attribute	Sub-attributes						
Molecule	Larger than[(1)]	Smaller than[(2)]	Struct. X	Struct. Y	Struct. Z	Struct. W	Struct. Q
Radiation	Radiation 700 nm	Radiation 450 nm	Radiation UV				
Global regulator							

concepts. As already mentioned, each attribute defined here corresponds to receptive ability of an organism to perform the following chain of actions:

- Recognise some stimulus (to form concept),
- Assimilate it, immobilise it and/or perform configuration change of itself, and
- Activate (indirectly or directly) some other molecules in order to proceed with processing of the received information.

3. Results and Discussion

Since this formal framework is unable to show sequential dynamics of some process we will construct several different formal contexts, and corresponding lattices, which will demonstrate how internal network responds to external stimuli.

The first formal context that is defined shows the initial steps in interaction with the environment. According to the previously described process, the living system has a limited set of active receptors and through them it determines which kind of external objects can be visible, and consequently, can be functionally treated. In this example (Table 2), the

Table 2. Formal context which represents initial steps in interaction with the environment

Object	Mol.	Rad.	Mol. larger than$^{(1)}$	Mol. small. than$^{(2)}$	Mol. strX	Mol. strY	Mol. strZ	Rad. 700 nm	Rad. 450 nm	Rad. UV
0		X						X		
1		X							X	
2		X								X
3	X			X						
4	X			X						
5	X			X						
6	X		X		X					
7	X		X			X				
8	X		X				X			

living system can initially sense water and small ions (denoted as Mol. small. than[(1)]), three kinds of organic molecular structures (denoted as X, Y and Z), and three kinds of radiation sources (denoted as 700 nm, 450 nm, and UV).

As it can be seen in Figure 1, the complete environment is divided into two main concepts which further diverge into several more specific ones. Since each formed concept in this model is connected with previously defined chain of events, in Figure 1b is shown how simple feedback governed metabolic chain would look like from the FCA perspective.

The association of molecules with structures Z (Mol strZ) with corresponding receptor (concept <10>), activates production of next molecule (<12>), which in turn generates the product (<13>). Generation of the final product inhibits production of strZ receptors, the intermediate product spontaneously degrades and some other metabolic chain utilises this final product (<13>). Since (<13>) is no longer available, production of strZ receptors can start again.

In the second formal context we introduced the activation of the supposed global regulator. As it can be clearly seen, complete lattice is reconfigured, and number, as well as hierarchy between concepts is drastically changed. It is very important to emphasise that activation of global regulators is the inevitable consequence of assimilation of almost any external stimulus. More generally, an interaction with the environment through building of subjective perspective, in case of living organisms, is connected with temporality of internal conceptual lattice. This means that finalising formation of one configuration at the same time causes its degradation and reconfiguration.

4. Conclusion

It is obvious that this model is an over-simplified example of a photosynthetic organism, but even in that form it can reveal some important characteristics of living systems. First, since they are able to form subjective perspective, living systems operate only within limited scope of the wholeness of the environment. Moreover, they shape the

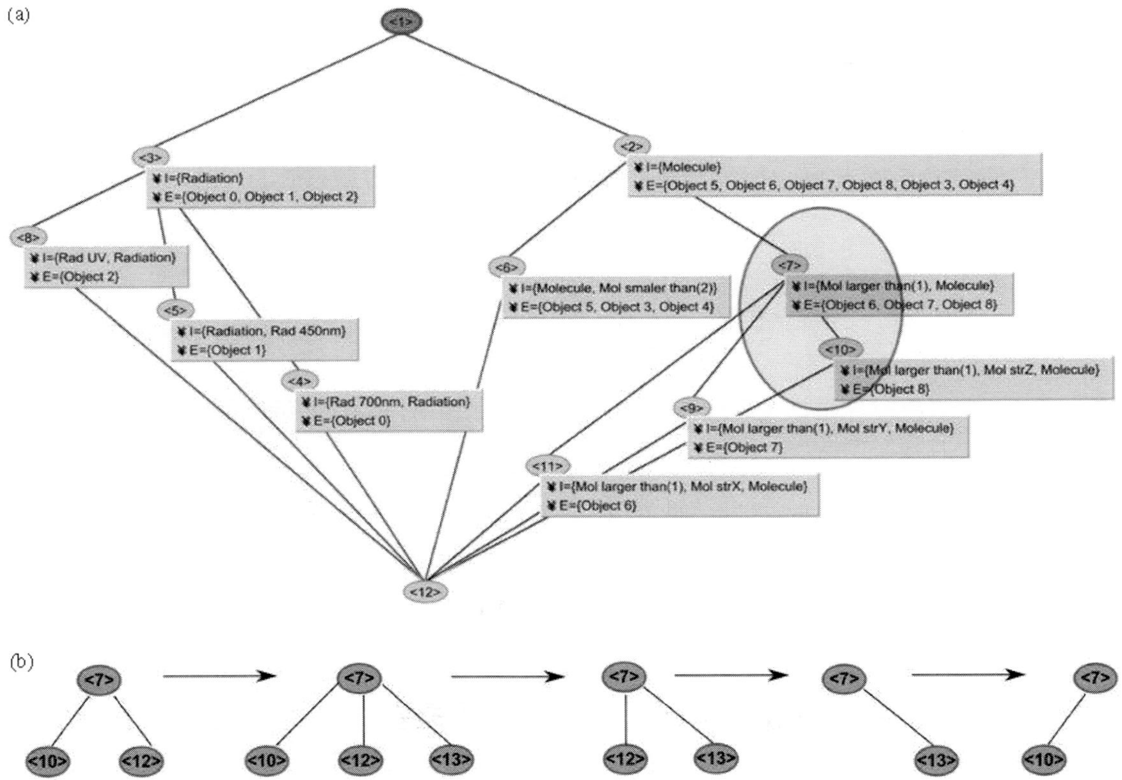

Figure 1. Lattice of the formal context given in Table 2 (lattice generation was performed using open-source software package "Galicia"[5]).

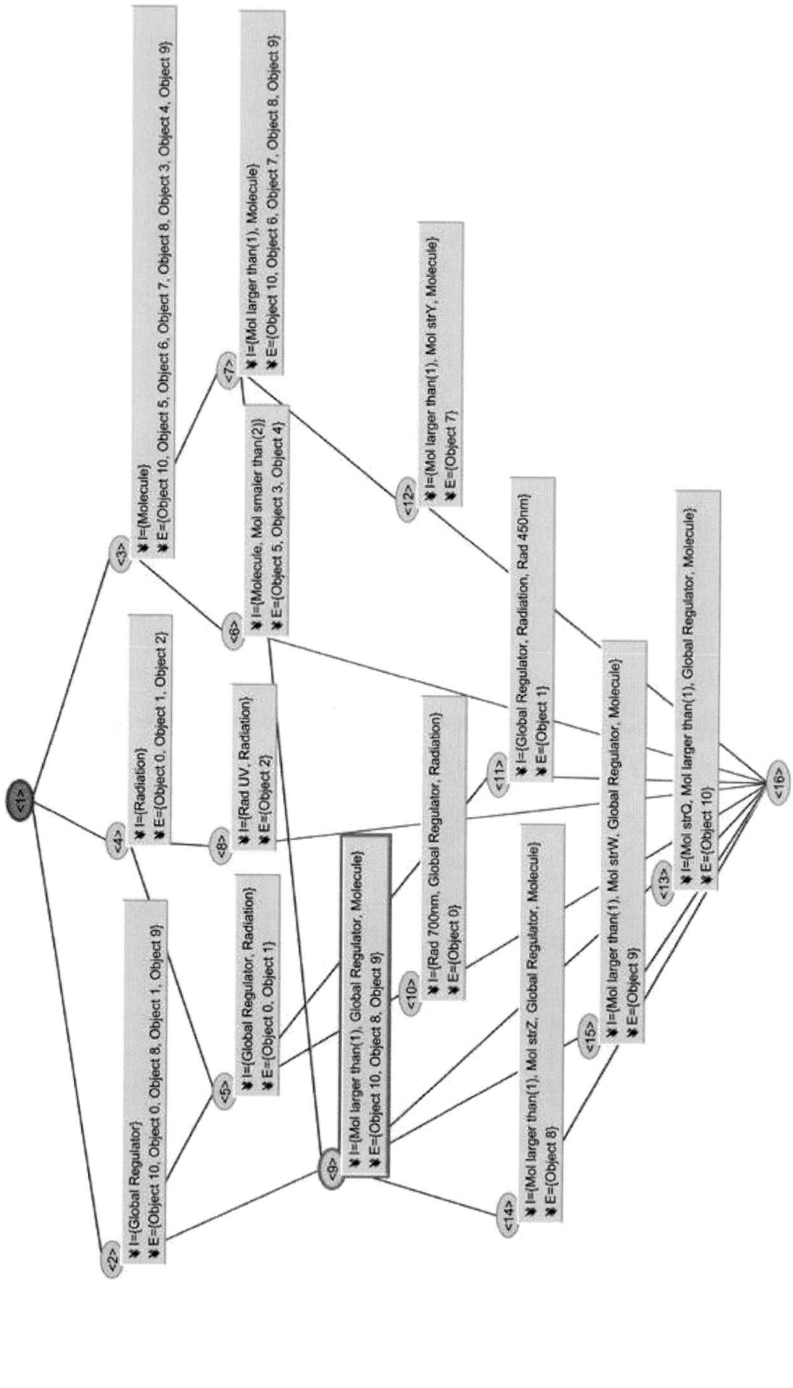

Figure 2. Lattice of the formal context for the same living system, when global regulator is active.

environment in accordance with their ability to functionally process some of external segments thus forming the subjective environment. This subjective construction is very changeable because it is defined not as a reflection of complete ability of certain living system to sense its environment, but only as a reflection of currently active receptors. Second, the formation of internal conceptual lattice is temporalised by the very same process of its formation. When it is formed, its consequence is activation of other molecules/receptors/regulators which have their perspective, thus changing the current context and reconfiguring almost the complete lattice.

References

1. O.E. Rössler, *Endophysics: The World as an Interface* (World Scientific Pub. Co. 1998).
2. N. Luhmann, Soziale Systeme. Grundriss einer allgemeinen Theorie, Publishing Company of Zoran Stojanović (1999).
3. R. Wille, Restructuring lattice theory: An approach based on hierarchies of concepts, Ordered sets, 445 (1982).
4. B. Ganter and R. Wille, *Formal Concept Analysis: Mathematical Foundations* (Springer-Verlag 1997).
5. Galicia, http://www.iro.umontreal.ca/~galicia/ (2001).

[illegible]

CHAPTER 22

AN EXAMPLE OF METEOROLOGICAL FIELD FORECAST OBTAINED BY THE SINGLE LEVEL NESTING FOR USE IN ECOLOGICAL MODELLING

ANA ĆIRIŠAN AND ALBERT RUMAN

ACIMSI — Centre for Meteorology and Environmental Modelling, University of Novi Sad, Novi Sad, Serbia

It is a common assumption that very high resolution model will produce more accurate prediction. Unfortunately, that requires large computing recourses since, for instance, doubling of horizontal resolution increases the computing time by eight times. One possibility is to perform two integrations. The first one for a larger domain with coarser resolution and the second one, which is nested inside the first, with much finer resolution, but accordingly to smaller domain. From the ecological point of view, more accurate prediction of meteorological fields like wind, temperature, humidity profiles and precipitation would improve the agro-meteorological, ecological and chemical modelling of different transport processes in the atmosphere. The comparison of model results using the Workstation Eta Model for the original grid and nested one is analysed in this paper. Verification of both forecasts was done against the available synop data.

1. Introduction

In order to improve forecasting of meteorological parameters, a number of numerical techniques are applied in numerical weather prediction models. Weather forecasting based on nested numerical models can significantly improve short and medium term weather forecasting.

An increase of accuracy and local specificity of the prediction of meteorological parameters are of great interest for agriculture and ecology in general. Also, the advances in quality and detail of short-term weather forecasting are very important for the air quality modelling. In recent years, the use of three-dimensional high frequency mesoscale data sets derived from dynamical models to drive air quality simulation models has grown. Researchers can also link numerical models with a chemistry model to track the formation and movement of chemicals, aerosols and particulate matter.

The aim of this work was to determine the impact of horizontal resolution and initial and boundary conditions of standard and nested models on environmental prediction. Numerical weather prediction model used in this paper is Workstation Eta model, a complete, full physics system nearly identical to the operational Eta model. Verification of the model forecast results obtained by coarser and finer resolution was done using the synop data.

2. Model Characteristic

Workstation Eta is a complete, full-physics, NWP system, developed at the National Centre for Environmental Prediction (NCEP). It is the research version of NCEP's operational Eta model. Workstation Eta is a regional model for numerical forecast of atmosphere conditions, for processes of synoptic or sub-synoptic proportions, giving the forecast two or three days forward.

The Eta model was originally developed from co-operation among the Institute of Meteorology of the University of Belgrade, GFDL Princeton and NMC Washington by Zaviša Janjić and Fedor Mesinger, with the specific objective of being applicable to regions with steep mountains.[16] It was named after its vertical co-ordinate η (eta). This co-ordinate is also known as the step-mountain co-ordinate because the surfaces are represented in the form of steps whose tops coincide with model layer interfaces. Eta co-ordinate is created to reduce errors that appear while calculating the pressure gradient force, advection and lateral diffusion along a steeply sloped co-ordinate surface. Opposite to the sigma system, which can be steeply sloped, co-ordinate surfaces of eta co-ordinate are approximately horizontal.

2.1. *Model Dynamics*

The Workstation Eta model has been available as a research and operational forecasting model since 1999. It is designed as a hydrostatic model (with a non-hydrostatic option), using the primitive equations based on the hydrostatic approximation.[15] It is a grid-point model with limited area integration domain. The semi-staggered Arakawa E grid[1] is the basis of the model's horizontal structure, see Figure 1. This grid staggers the mass variables and the wind variables. *H* represents a "mass" variable point (temperature, pressure, moisture); *V* represents both horizontal components of the wind. The magnitude of the distance between the two nearest grid points carrying the same variable indicates the model's horizontal resolution.

Time differencing is performed with split-explicit time differencing schemes,[4,5,11,12] providing higher model computational efficiency. Horizontal advection scheme is used that controls energy cascade toward smaller scales for the E grid.[6] The mountains in the eta system are represented as three-dimensional "boxes" on grid points.[14] Proper

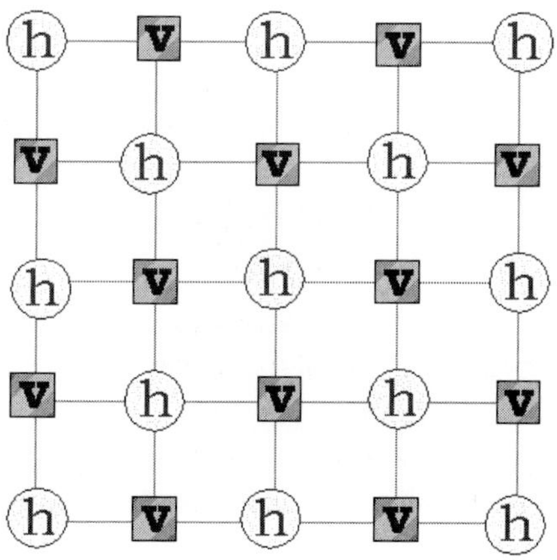

Figure 1. The variables distribution on the E grid.

representation of the mountain effects was the basic reason to introduce the eta vertical co-ordinate system. Vertical, eta (η) co-ordinate[13] is defined as:

$$\eta = \frac{p - p_T}{p_S - p_T}\eta_S \tag{1}$$

where

$$\eta_S = \frac{p_{rf}(z_S) - p_T}{p_{rf}(0) - p_T} \tag{2}$$

p is a pressure, the subscripts T and S represent values on the top of model's atmosphere and on the surface, z is height, and $p_{rf}(z)$ is a suitably defined reference pressure as a function of z. The ground surface heights z_S are permitted to take only a discrete set of values, so $p_{rf}(0)$ can be replaced by $p_{rf}(z_b)$, where z_b represents basic height defined so as to include gentle slopes of the surface topography. Figure 2 shows mountains and their vertical discretisation with the η co-ordinate. For each model grid box, T represents "mass" variables

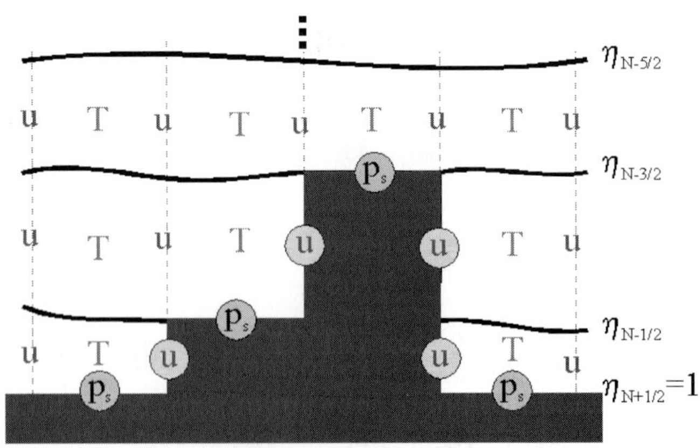

Figure 2. Distribution of vertical levels in the model with explanation of mountain discretisation.

such as temperature and moisture, while u represents both horizontal components of the wind and p_S is the surface pressure. The circled u grid points would be the points with zero values of the velocity components normal to sides of the mountain shown in the figure.

2.2. *Model Physics*

The physical part of the WS Eta model is based on several sophisticated parameterisation schemes. Vertical turbulent mixing between levels in the free atmosphere is performed by using mixing coefficients of the Mellor-Yamada 2.5 level turbulence.[7,10] Vertical mixing in the surface layer is performed by a Monin-Obukhov similarity model. Nonlinear fourth-order lateral diffusion scheme is used with diffusion coefficient that depends on flowing field deformation and turbulent kinetic energy introduced in order to control the level of small-scale noise. The default cumulus parameterisation scheme is the Betts-Miller-Janjić (BMJ) deep and shallow cumulus convection scheme, used in order to represent moisture processes responsible for excessive precipitation events,[2,8] computed every 16 adjustment time steps. Also, the Ferrier Eta microphysics is included. The GFDL (Geophysical Fluid Dynamics Laboratory) radiation scheme is used, which includes the short wave parameterisation of,[9] two-stream multi-band scheme with ozone from climatology and cloud effects and the long wave parameterisation of,[3] an older multi-band scheme with carbon dioxide, ozone and microphysics effects. The Oregon State University (OSU) scheme for prediction of the surface processes, including surface hydrology is incorporated in the model.

2.3. *Initial and Boundary Conditions*

The Workstation Eta is used as a limited area model so initial and boundary conditions are necessary to perform simulations. The input data for the model are derived from the National Centres for Environmental Prediction (Washington DC). NCEP-GFS grib files with 1-degree spatial resolution are used to create the necessary initial and boundary conditions updated at 6-h intervals, giving 6-hourly forecast data. The initial fields for decoding are analysis of American meteorological service, files in Gridded Binary (GRIB) format, a compressed format used for gridded meteorological data. Model grid sizes are variable and flexible, depending on the scientific problem and the computing resources available. The present operational version of the Workstation Eta model has a vertical resolution with 38 levels with a pressure level spacing of 25 hpa from the surface to the top model atmospheric level of 25 hpa. The model gives predictions up to 72 hours (3 days) and resolutions are chosen to optimise user requirements and the computational resources and the desired quality of the forecast output. The model's boundary data are obtained by direct linear interpolation from the AVN run of the Global Model. The mountains in a model are formed from NCEP global topography, with 30 × 30 sec of degree resolution that includes approximately 600 m × 600 m area for middle latitudes. The model point height is an average of all points in a square dx × dy. Vertical discretisation defines the mountain height in the model. Soil and vegetation types are defined at 1 km resolution.

After data decoding and making the topography, it is necessary to make bilinear interpolation of data from longitude-latitude grid into the model grid. Also, it is necessary to make vertical interpolation from standard pressure levels into eta co-ordinate surfaces. This interpolation uses pressure alogarithm. After the model run, a postprocessor is used to map the forecast output data from the native Eta grid to the latlon grid for use by the graphics package. Fortran Power Station programming package is used for extraction and data inscription. The visualization software used to display the model output includes Grid Analysis and Display System (GraDS) and Origin Software.

3. Nesting Capabilities

In order to get a more precise forecast, a nested model has been used. The nesting procedure allows a very fine model to be set up on a limited area, using results from a coarser grid simulation to impose the boundary conditions at the open boundaries of the finer model. The nested model actually involves two simulations running simultaneously. The main grid can be relatively coarse. This domain is run first. Periodical outputs are stored and then used to constrain the run of a nested grid. The secondary model (nested one) is set up on a much smaller area. It is a higher-resolution model run, which gets initial and lateral boundary conditions from a previous run (the main model). The conditions at those boundaries will be given by interpolating results from the main model at each time step.

4. Results

Based on these considerations the following domains and resolutions have been implemented. The simulations are initialized at 00 UTC 03 Jun and run through 00 UTC 06 Jun (72 h) 2007. Our downscaling strategy was to move from 60 km model-scale and 25 km model-scale to 2 km nested one. The centre of domain is 9.4°E, 50.6°N. The grid resolution is 0.577 degrees in longitude and 0.538 degrees in latitude which approximately translates to 64 km × 59 km horizontal grid. The time step for this resolution is set to 200s. Second model run is configured to 25 km × 23 km grid spacing (grid resolution is 0.222 degrees in longitude and 0.205 degrees in latitude) with a 71 × 71 number of grid points and with time step of 90 s. Two more runs are performed with the same horizontal grid, but with another grid domain. First domain covers the grid domain of 41 × 41 grid points. Number of grid points used in second case was 151 × 151. Then, a high resolution terrain was incorporated. The central point for nested model domain is 20°E, 45°N. The 2 km horizontal resolution model is run using the one-way nesting method. For every low-resolution model mentioned above a single level nesting is done. Initial and lateral boundary conditions are determined (interpolated) from the outer models. The inner nest is run using grid resolution of 0.02 degrees in longitude and 0.0198 degrees in latitude with 31 × 51 grid points. The time step required by the stability criterion for this resolution is set to 6 s. The largest model is also nested on 7 km horizontal resolution, using grid resolution of 0.067 degrees in longitude and 0.066 degrees in latitude and time step of 20 s. Computations are done for various model fields including: surface pressure, temperature, water vapour, cloud water, horizontal wind (u, v), several surface/soil parameters, vertical velocity, geopotential height, precipitation, etc.

Model variables of great interest for ecological modelling analysed in this paper are shelter temperature, and wind course and intensity. All model outputs are compared with synop data at 6 h interval.

As we can see from Figure 3 and Figure 4, both kinds of resolutions show a good correlation with the observed data on the first day of forecast, low resolution models even better than nested ones. Within the next 48 h low resolution models have greater bias from the observed data than high resolution models. Also, as we can see, nesting from the larger domain gives inferior outputs compared to outputs from high resolution models, nested within a coarser resolution with double smaller domain.

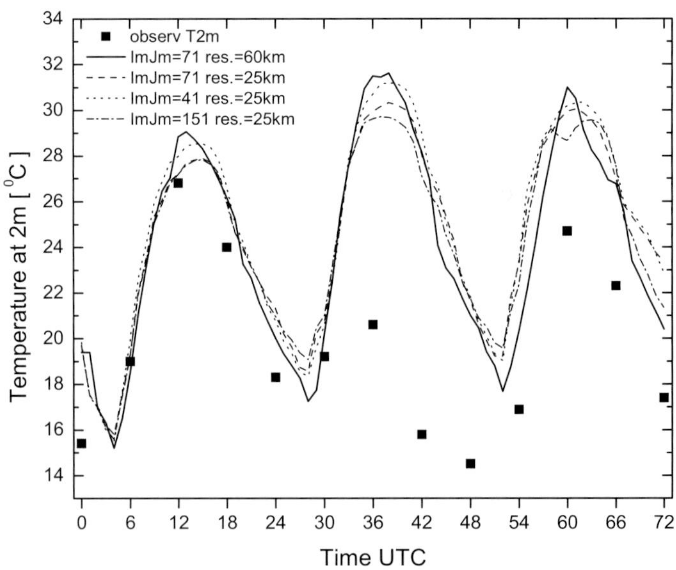

Figure 3. Low-resolution model outputs versus observed data for temperature at 2 m.

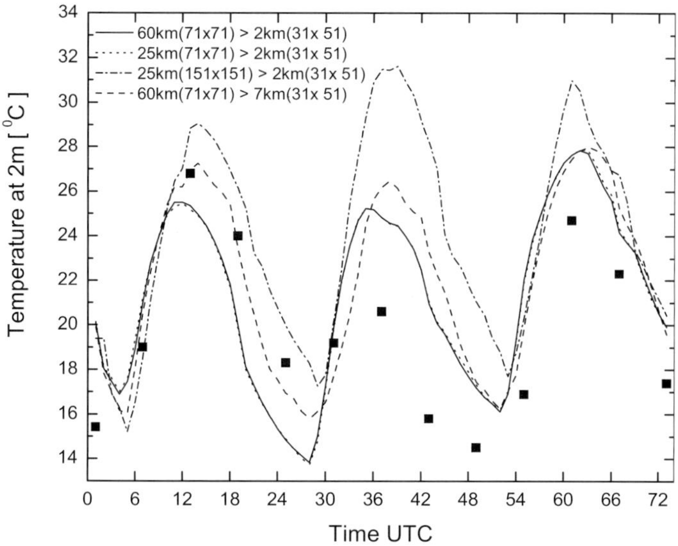

Figure 4. High-resolution model outputs versus observed data for temperature at 2 m.

In Figures 5–10, the results of wind course and intensity obtained by models outputs are compared versus the observed data. Within the models of coarser resolution there is a quite well accompaniment through all the period of forecast run. In case of nested run, see Figure 6, Figure 8 and Figure 10, better correlation with the observed data is during the first 24 h of wind forecast.

There is a great impact of meteorological variables on ecological modelling. The outputs of such a model as WS Eta may be used as inputs to further modelling. Since most air quality modelling applications involve fine grid resolution, fields from the fine-scale nested grids are used as the input to air quality models.

High resolution winds from NWP models will allow more accurate representations of near-surface flows. Detailed modelling of the wind is important for prediction of the trajectory-directions. Also, it is important for accurate simulations of the transport, diffusion, dry and wet deposition, and chemical transformations for aerosols and particulate matter.

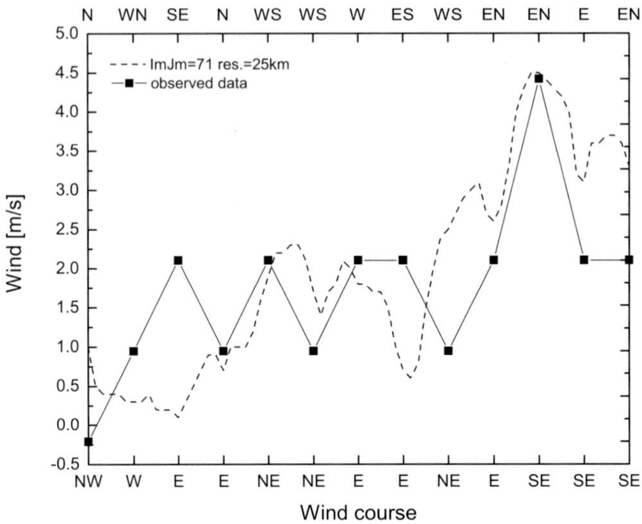

Figure 5. Low-resolution model output (25 km horizontal resolution with 71 × 71 grid points) versus observed data for wind course and intensity.

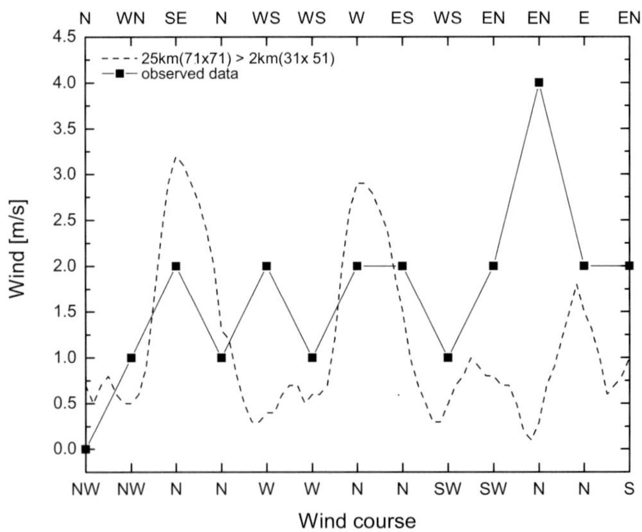

Figure 6. High-resolution model output (2 km horizontal resolution with 31 × 51 grid points nested from model with 25 km horizontal resolution and 71 × 71 grid points) versus observed data for wind course and intensity.

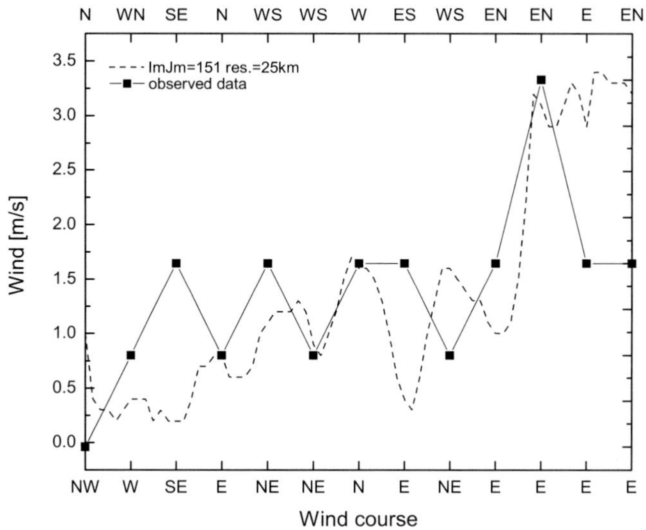

Figure 7. Low-resolution model output (25 km horizontal resolution with 151 × 151 grid points) versus observed data for wind course and intensity.

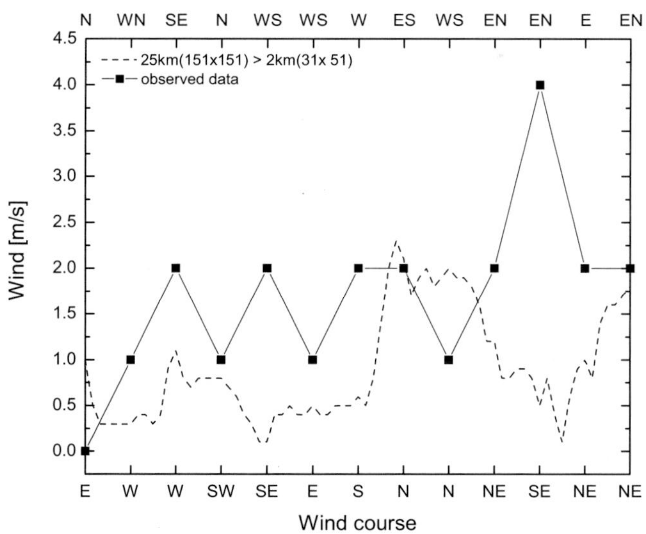

Figure 8. High-resolution model output (2 km horizontal resolution with 31 × 51 grid points nested from model with 25 km horizontal resolution and 151 × 151 grid points) versus observed data for wind course and intensity.

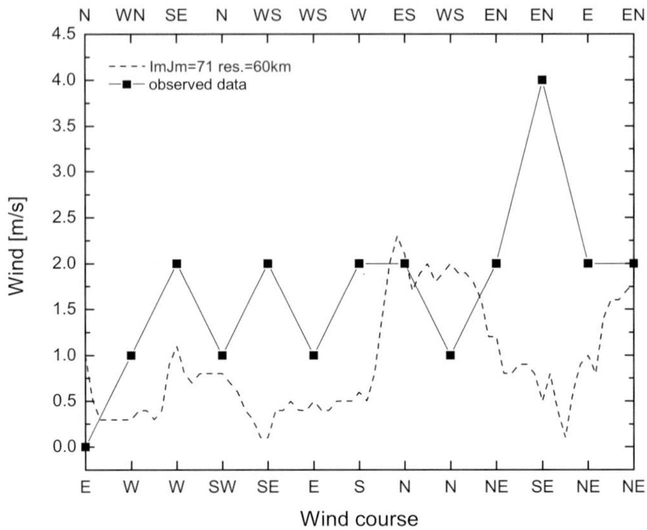

Figure 9. Low-resolution model output (60 km horizontal resolution with 71 × 71 grid points) versus observed data for wind course and intensity.

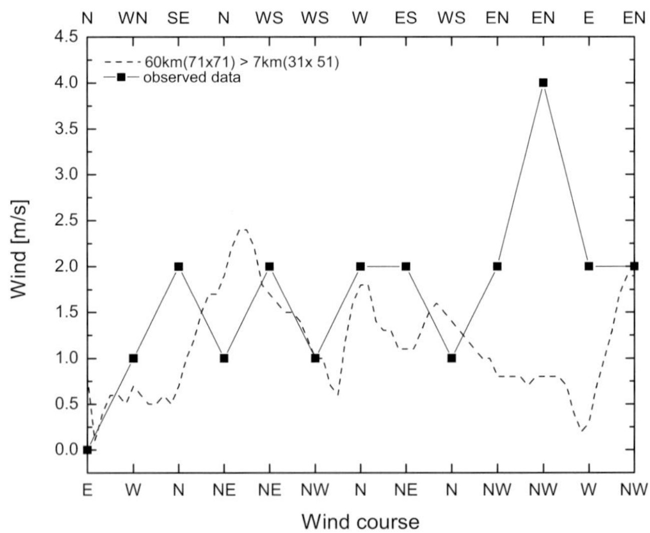

Figure 10. High-resolution model output (7 km horizontal resolution with 31 × 51 grid points nested from model with 60 km horizontal resolution and 71 × 71 grid points) versus observed data for wind course and intensity.

Higher temperatures increase transpiration at the leaf surface and lead to an accelerated ageing of the foliage, and a shortening of the growing season. But it should also be realised that the result of a higher ambient temperature, leading to a higher evaporative demand could be overridden by changes in other environmental and atmospheric conditions like soil water availability, precipitation patterns, cloudiness and air humidity. Changes in precipitation distribution may cause changes in soil moisture and, combined with changes in (air and soil) temperatures, markedly change soil fertility and thereby local farming systems and crop productivity.[17]

The fire model can take its initial and boundary conditions from the finest grids used in model. The coupled atmosphere-fire model will simulate fire spread, heat, moisture and particulate (smoke) exchanges.

5. Conclusion

There is a great ability of ecological models to interface with a variety of meteorological models. Downscaling from global to regional application should be implemented due to the nature of the problem to be solved. To determine the relative influence of the Workstation Eta model initial and boundary conditions on the circulations of great interest for ecological models, several simulations were done. Advantage of high resolution models is their efficiency opposite to the low resolution ones. In order to determine the accuracy of those models it is necessary to run more simulations, defining for which meteorological fields nesting is recommendable or not.

References

1. A. Arakawa and V.R. Lamb, Computational design of the basic dynamical processes of the UCLA general circulation model, *Methods in Computational Physics*, J. Academic Press, New York, 173 (1977).
2. A.K. Betts and M.J. Miller, *Quart. J. Roy. Meteor. Soc.* **112**, 693 (1986).
3. S.B. Fels and M.D. Schwartzkopf, *J. Atmos Sci.* **32**, 1475 (1975).
4. Z.I. Janjic, *Mon. Wea. Rev.* **102**, 319 (1974).

5. Z.I. Janjic, *Contrib. Atmos. Phys.* **52**, 69 (1979).
6. Z.I. Janjic, *Mon. Wea. Rev.* **112**, 1234 (1984).
7. Z.I. Janjic, *Mon. Wea. Rev.* **118**, 1429 (1990).
8. Z.I. Janjic, *Mon. Wea. Rev.* **122**, 927 (1994).
9. A.A. Lacis and J.E. Hansen, *J. Atmos. Sci.* **31**, 118 (1974).
10. G.L. Mellor and T. Yamada, *J. Atmos. Sci.* **31**, 1791 (1974).
11. F. Mesinger, *Tellus.* **5**, 444 (1973).
12. F. Mesinger, *Beitr. Phys. Atmos.* **50**, 200 (1977).
13. F. Mesinger, *Riv. Meteor. Aeronautica.* **44**, 195 (1984).
14. F. Mesinger, Z.I. Janjic and S. Nickovic, D. Gavrilov and D.G. Deaven, *Mon. Wea. Rev.* **116**, 493 (1988).
15. S. Nickovic, D. Jovic and G. Kallos, *Proceedings of the 3rd Panhellenic Conference on Meteorology, Climatology and Atmospheric Physics*, Athens, Greece, 45 (1996).
16. E.P. Mathew, A guide to the Workstation ETA, *NCEP Publication*, Washington, 17 (2000).
17. S.C. Van de Geijn and J. Goudriaan, *Global Climate Change and Agricultural Production*, eds. F. Bazzas and W. Sombroek, Vol. 101 (FAO and John Wiley & Sons, 1996).

CHAPTER 23

USE OF THE SPATIAL ANALOGY TO UNDERSTAND THE EFFECTS OF CLIMATE CHANGE

LEVENTE HORVÁTH AND NORBERT SOLYMOSI

MTA-BCE Adaptation for Climate Change Research Group, Budapest, Hungary

MÁRTA GAÁL

Faculty of Mathematics and Informatics, Corvinus University of Budapest, Budapest, Hungary

Several climate models were elaborated to predict the climate change tendencies. This work is based on the HadCM3 model. Our target (reference) site was Debrecen, an important centre of agricultural production in Hungary. Scenario values for Debrecen were calculated from the TYN SC 1.0 10 minute resolution grid climatic database, in different time periods. Then the method of spatial analogies was used to understand the meaning of these scenario values. Spatial (geographical) analogues are regions which today have a climate analogous to that predicted in the study region in the future. It has been found that regions similar to the predicted future climate of Debrecen are located south to Hungary. At the closest time this distance is about 100–300 km, but later the scenarios predict more drastic change. In the period 2011 — 40 all the scenarios (A1FI, A2, B1, B2) are similar, they predict mostly the same, and differences are only later in the 2041–70s. If we accept the result of spatial analogy, it seems that in the near future the climate of Debrecen and its region will be similar to North Serbia (Vojvodina region) or South Romania. We show the results of the A1FI scenario.

1. Introduction

Nowadays we often hear about the climate change, but the meaning of this word is different for the ordinary people and for the scientists. We can feel in everyday life that the weather is not such as it used to be, the extremity is growing, heat waves and big precipitation become more often. To understand the climate change and its effects we have to translate it to mathematical language. We have to use the results of the Global Climate Models (GCMs), to examine how the world's climate will change. The models are showing us only possible pictures of the future, we cannot determine exactly what will be in the next year, after 10 years or at the end of the century. The job of the scientist is to transform the result into clear, understandable form. One of the methods to understand these results is the analogy. Analogies have three main types:

- The temporal analogy, which is used by the paleoclimatology. It is using measured climatology, to predict the future climate. Unfortunately it is working on long time periods.
- The incremental analogy, which is using a special method, it is calculating the future climate by using the observed climate, and changing it additively.
- The third is the spatial analogy.

It is using the results of the GCMs. With this method we are searching for regions which at present have the same climate as the scenarios indicated for the future. With the spatial analogy we can make the climate scenarios easier to understand. In this research the analogue regions for Debrecen, which is an agriculturally important region for Hungary, were examined. From the result we can say — according to international results — that our climate will be similar to regions south to Hungary. This shifting will be 100–300 km in the next decades (2011–2040), and it could be 400–500 km in the middle of the century, and maybe there no spatial analogues in Europe to the end of the century. Different methods were used to calculate the analogues, but they indicate the same regions. These analogue regions are North-Serbia, Vojvodina regions and South Romania.

2. Materials End Methods

2.1. *Climatic Data*

Baseline climate data of Debrecen were taken from the OMSZ (Hungarian Meteorological Office) database. For the European base climatology (1961–1990) the TYN CL 2.0[1] database was used, which contains 10 minutes spatial resolution grid data for 5 climatic parameters. From these the monthly temperature and precipitation data were used, only. We used the well-known, internationally used emission scenarios — A1, A2, B1, B2 — from the Hadley Centre's HadCM3 model. In this paper the results of the A1FI scenario are presented. For the scenarios, the TYN SC 1.0[2] database was used, and we calculated the monthly averages for the 2011–2040 time period.

2.2. *Methods*

For the calculation of spatial analogy, two methods were used, the Rank-method[3] and the CLIMEX method.[4] With the Rank-method for each grid boxes the Euclidean distances of the climatic parameters were calculated and ranked. This method shows only the closest points, the most similar grid's similarity is 100 the less similarity is 0. It means that we always have exclusively similar point, which may be not really similar. The other disadvantage of the method is that the similarity depends on the number of the examined grid points. With the CLIMEX method the similarity index for two climate datasets can be given, so that with parameter settings the user can define what he calls similar. The method's advantages are that does not depend on the number of grid points, and shows really the similarity.

3. Results

3.1. *Validation of the Two Methods*

For proofing that the methods are working well, the analogue regions for the base period were calculated. The method is appropriate if the spatial analogues of the examined region (Debrecen) are also in the examined region. In the maps regions with more than 90% similarity can be seen.

In the maps, (see Figure 1), it can be seen that the analogue regions are mostly at the surroundings of Debrecen. The Rank-method shows higher similarity than the CLIMEX method, this is because by ranking, 10% of the grid points always have more than 90% similarity.

The difference of the methods can be seen at the calculation for the scenarios, too, see Figure 2. They show similar spatial analogues, but the Rank-method gives greater analogue regions, and shows higher similarities. With this method, we probably also get back such regions, which have not really similar climate. For this reason the CLIMEX method was selected for further analyses.

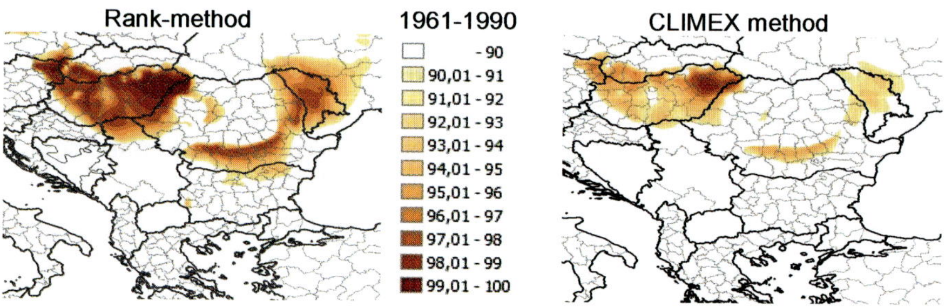

Figure 1. Analogue regions for the base period.

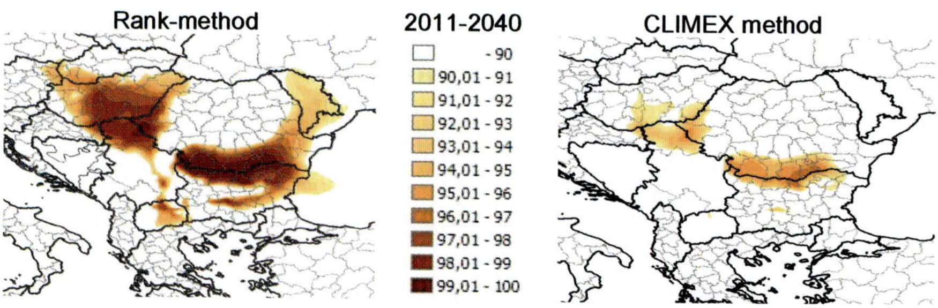

Figure 2. Spatial analogues for the A1FI scenario.

3.2. *Sensitivity Analysis*

The CLIMEX method basically applies the temperature and the precipitation data together, with the same weights of the distances. By the help of a λ parameter we can apply different weights for temperature and precipitation, which could be useful in some agroecological analyses. It is also useful to test sensitivity of the CLIMEX method. When $\lambda = 1$ it means that the analogues are calculated only based on temperature, when $\lambda = 0$ only based on precipitation. Further examined values of λ were 0.2, 0.4, 0.6 and 0.8.

The analogue regions obtained by the temperature ($\lambda = 1$) and precipitation ($\lambda = 0$) can be seen in Figure 3. Results show similarities with the previous maps (temperature and precipitation together in Figure 1 and Figure 2), in the base period give back the surroundings of Debrecen, but other areas also appear as analogues. The same tendency can be observed in case of the scenario. These regions

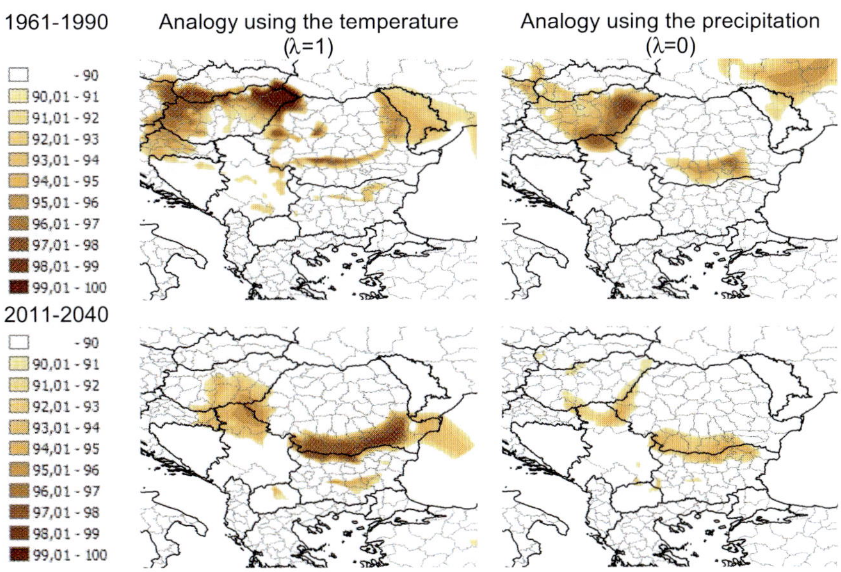

Figure 3. Analogue regions for the base and 2011–2040 by the temperature and the precipitation.

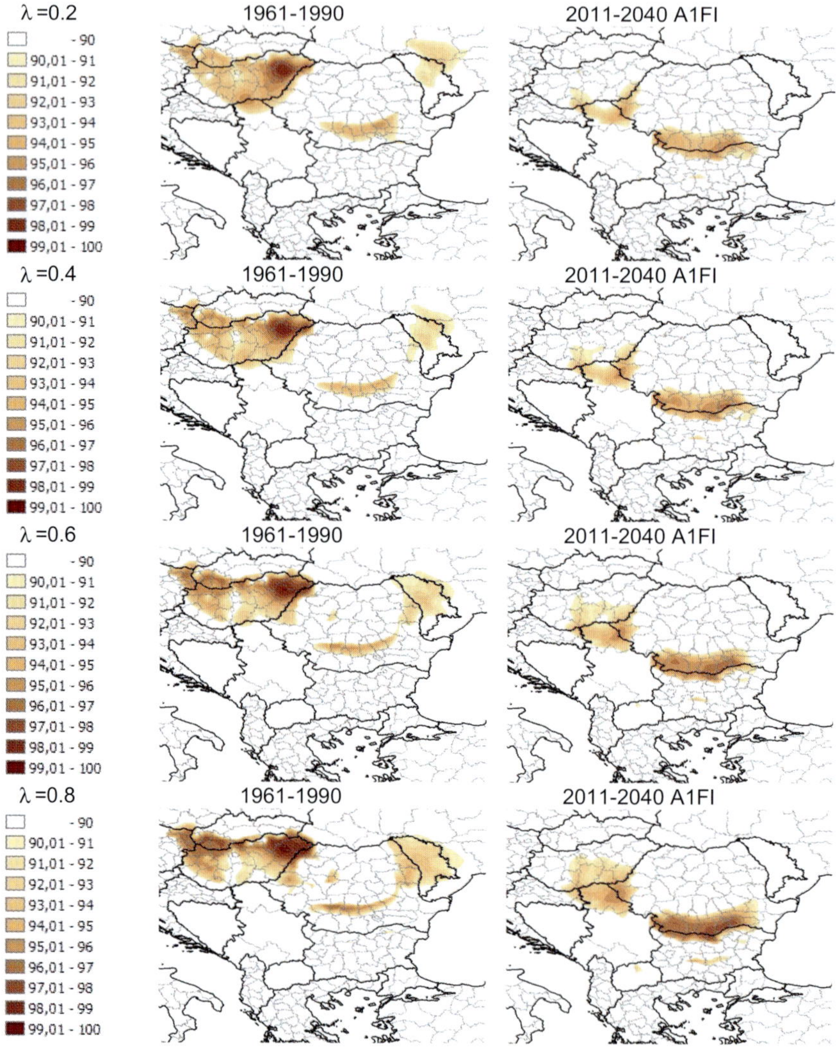

Figure 4. Analogue regions using different λ values.

can be analogues regarding temperature but not precipitation, or vice versa.

Results obtained using different λ values can be seen in Figure 4. It shows a gradation between the analogue regions for precipitation and temperature. The differences are not critical, it means that the method is not sensitive to the λ value.

But if the purpose of the analyses does not require the use of only one climatic parameter, we recommend using the CLIMEX method with the temperature and precipitation together, with the same weights.

We found that the method of the spatial analogy is a good tool to understand the effects of the climate change. Finding the analogue regions we can compare their cropping system, land use and other ecological and economic characteristics with the present ones in Hungary. It helps to develop adaptation and mitigation strategies on climate change.

4. Conclusion

The two examined methods — Rank-methods and CLIMEX — for finding the spatial analogue regions gave almost the same results. But we recommend using the CLIMEX method with the same weights of temperature and precipitation. If we accept the results of the GCMs, according to the A1FI scenario for the 2011–2040 periods, the analogue regions of Debrecen will be at Vojvodina region in Serbia, and South-Romania. It means about 100–300 km shifting to south, which correspond to other international results.

The detailed analyses of the analogue regions can help us to adapt to the changing climate. From the analogue regions we should collect all kinds of available ecological, agricultural, economical, social or public sanitation data. We can study what kind of problems are there, and what are the solutions. We can learn from there how to solve the possible problems and develop strategies. This will be a good base for further research and an important base for decision makers.

Acknowledgments

Our work was supported by OTKA T 042583 tender and Klima KKT project.

References

1. M. New, M. Hulme and P.D. Jones, J. Climate **12**, 829 (1999).
2. T.D. Mitchell, T.R. Carter, P.H. Jones, M. Hulme and M. New, *J. Climate* **13**, 2217 (2003).
3. E. De Pauw, An agroecological exploration of the Arabian Peninsula, ICARDA, Aleppo, Syria (2002).
4. A.M. Young, B. Blackshaw, G.F. Maywald and R.W. Sutherst, CLIMEX for Windows 1.1. Tutorials, CSIRO, Melbourne (1999).

CHAPTER 24

MODELLING TEMPERATURES IN CORN AS THE CONSEQUENCE OF GLOBAL CLIMATE CHANGE IN HUNGARY

TIMEA KOCSIS, ANGELA ANDA AND JUDIT BEM

Faculty of Agronomy, University of Pannonia Georgikon, Keszthely, Hungary

Local consequences of global climate modifications in corn canopy were studied at Keszthely (Hungary). All model inputs (meteorological and plant features) were collected at Keszthely Agrometeorological Research Station during the past three decades. We applied the crop microclimate simulation model (CMSM) of Goudriaan using 6 scenarios, with the first one containing the actual data. Other scenarios reflect several possible status of climate modifications predicted in the examined area. Out of model outputs, the crop- and inside canopy air space temperatures were the focal points of our study. We concluded that the temperature prognosis showed close connection not only with meteorological (environmental) elements, but also with actual canopy architecture, the leaf area index and density of leaves.

1. Introduction

The atmospheric CO_2 enrichment of anthropogenic origin leads to global warming as a part of climatic change. The global atmospheric concentration of carbon dioxide has increased from a pre-industrial value of about 280 ppm to 379 ppm in 2005. The annual fossil carbon dioxide emissions increased from an average of 6.4 GtC per year in the 1990s to 7.2 GtC per year in 2000–2005.[5] The effects of climate change are diverse, but they do not affect equally different parts of the Earth.

Table 1. Local changes of air temperature and precipitation at a certain level of global warming according to J. Mika[8]

Global Local change	+0.5°C	+1°C	+2°C	+4°C
Air temperature (°C) summer/summer half-year	+1.0°C	+1.3°C	+2°C	+4°C
Air temperature (°C) winter/winter half-year	+0.8°C	+1.7°C	+3°C	+6°C
Precipitation Annual amount	–40 mm	–66 mm	Uncertain	+40–400 mm

Prognostics of the impacts for the Carpathian-basin (Hungary) need regional climate scenarios of adequate time and space demolition, for concretising correctly the global alternatives to local scale. According to model evaluations, the local changes of air temperature and precipitation in the range of 0.5–4°C global change are described in Table 1. The main statement of the scenarios is that the local weather would get drier and richer in sunlight as the result of increasing greenhouse effect, at least in the first several decades of the global warming.[8] The relative sensitivity of the summer/summer half-year air temperature decreases from the twofold at the beginning to simple, while in the winter half-year the sensitivity remains at sesquialteral ratio.

The annual precipitation amount follows the warming non-linearly: the tendency of severe drying of the beginning phase turns to positive change in case of serious warming of 4°C.[8] Bartholy *et al.*[2] estimated the regional effects of climate change at the Lake Balaton-Sió Canal catchment area (West Hungary) by their own developed stochastic-dynamic downscaling model using the ECHAM/GCM outputs. (The place of our research was in Keszthely, situated on the west shore of the Lake Balaton.) They predicted a decrease of 25–35% of precipitation amount in the summer half-year and 0–10% decline in the winter half-year at a climate corresponding to a doubled CO_2 level.

The soil–plant parameters are needed in atmospheric, hydrological or ecological models. They require accurate description of the surface energy and water fluxes to assure the realistic simulation of partitioning the available energy into sensible and latent heat fluxes as well as the cycling of carbon through different organic and inorganic phases.[6]

Our aim was to investigate how locally grown corn reacts to possible climate change at the level of plant stands temperatures. Temperature regimes (crop air space temperature, crop foliage temperature) within the crop environment strongly influence its growth, diseases and yield. They may be identified as the key environment variables strongly related to the crop yield, phenological development, intensity of evapotranspiration, leaf water potential, water status, etc.[7]

For the analysis we created scenarios of probable climate statuses according to the above-mentioned references. For our investigation we applied the Crop Micrometeorological Simulation Model developed by Goudriaan[4] and reworked by Chen.[3]

2. Material and Methods

2.1. *The Place of the Study with Data Collection*

In our study we applied the modelling tool, one of the possibilities in assessing the future consequences of climate modifications on corn stand. The aim of the investigation was to forecast changes in temperatures of locally grown corn canopies. The inputs, both meteorological and plant features used in the simulation were collected at the Keszthely Agrometeorological Research Station (LAT: N 46°44′, LON: E 17°14’, ALT: 114.2 m). Although daily meteorological data were available from 1973 (the Agrometeorological Research Station was established in 1972), the data-set of the last ten years were used, because a QLC-50 automatic climate station was established in 1996. The sampling frequency of the automatic station (every six seconds) over the last ten years allowed the hourly means to be applicable as model inputs. Hourly averages were not available previously for the whole observation period, since the observations were made every 6 hours. Our sample day was an average day in July, when the plants were fully developed.

The reference level of the model inputs was taken into account by calculated aerodynamic depths for every stage of the plant development.[4] We estimated wind speed at a reference level using a combination of friction velocity and logarithmic wind speed profile above the canopy.[4] In case of the wind speed, measurements were made at 10 m above the ground.

The inputs of the model are site- and plant-specific values (plant height, leaf density in different layers), soil characteristics and hourly meteorological data (air temperature, global radiation, relative humidity, soil temperatures in different depths at 00.00 hours), which were transformed from the standard measurement level to the reference level required by the model (see above). The leaf area and its density were measured in the field on the same 10 sample plants weekly, using a LI-3000A type leaf area meter. In the present case the results of cob layer in the fully developed stand were presented (July). Regarding soil moisture, the monthly average for July of the past decade (–7 bar of soil water potential) was applied. The soil moisture content in the upper 1 m was also measured in the field gravimetrically every 10 days at 10 cm intervals.

2.2. *Brief Description of the Model*

The theoretical background of the Crop Micrometeorological Simulation Model (CMSM) is the physics of energy transformation and of transport processes. The model evaluates the parameters of micro-environment and plant characteristics by the principles of soil-atmosphere physics and plant physiology.[9] The basis of the functioning models follows the way of the incident solar radiation in different canopy layers. One part of the incident radiation is reflected from the canopy, the second part is transmitted by the layer and the third part is absorbed by the plants.

The model provides the detailed calculation of profiles of meteorological elements within the plant stand. It consists of three main parts: radiation-, aerodynamical-, and soil sub-model. The CMSM is a multi-layer steady state model, according to the air conditions, assuming equilibrium at any time in the air, and dynamic for the soil heat transfer because of the significant value of soil heat capacity.[4]

The theory of CMSM is the calculation of the radiation distribution among different environmental processes. The sensible heat flux (H_i) [J m^{-2}] in the ith layer is:

$$H_i = \frac{(T_{L,i} - T_{a,i})\rho c_p}{r_{H,i}} \tag{1}$$

where $\rho\ c_p$ is the volumetric heat capacity of the air [J m^{-3} K^{-1}]
$T_{L,i}$ is the temperature of the plant [°C]
$T_{a,i}$ is the air temperature [°C]
$r_{H,i}$ is the resistance against heat transmission [s m^{-1}].

The latent heat flux (λE_i) [J m^{-2}] in the ith layer can be calculated as follows:

$$H_i = \frac{(T_{L,i} - T_{a,i})\rho c_p}{r_{H,i}} \tag{2}$$

where γ is the psychrometric constant [mbar K^{-1}]
$e_{s,Ti,j}$: is the saturation water vapour pressure at actual plant temperature [mbar]
$e_{a,i}$ is the vapour pressure of the air [mbar]
$r_{V,i}$ is the resistance against the entrance of moisture into the layer [s m^{-1}].

After calculation of sensible and latent heat, the estimation of the air temperature in the ith layer ($T_{a,i}$) [°C] was as follows:

$$T_{a,i} = T_{a,i-1} + H_i \frac{R_i}{\rho c_p} \tag{3}$$

where R_i [s m^{-1}] is the value characteristic to resistance in the ith layer.

The model estimated the crop temperature ($T_{L,i}$) [°C] in the ith layer according to the following equitation:

$$T_{L,i} = T_{a,i} + (H_i - H_{i-1}) \frac{r_{H,i}}{\rho c_p} \tag{4}$$

From the *outputs*, i.e. crop temperature and air temperature inside the canopy, the air space was presented on the border of the upper third of plant height in the study. This is the place of cob formation, where the intensity of physiological processes is the highest.

2.3. *The Applied Scenarios*

In our study we established scenarios that represent climatic variations for Hungary. In the last two decades — in accordance with a tendency — the amount of rainfall decreased at Keszthely (10–15%). Based on the majority of publications regarding local climate modifications this trend is to be continued in the future together with air temperature increases (1.3–2°C in the summer half-year). The inputs of plant architecture, assimilatory surface size and its density were taken from the local measurements during the past three decades using the principle of analogy. We used plant data from the seasons in which air temperatures and soil water contents were the same in July as in our scenario:

- Control: It represents the present climatic conditions (average July day), average soil moisture content in this period and 380 ppmv ambient CO_2 concentrations. The value of LAI was 3.0, the characteristic mean for corn grown at Keszthely.
- Scenario 1: Ambient CO_2 concentration was doubled together with reduction of soil water content by 25%. In addition we reduced LAI to 2.6.
- Scenario 2: Ambient CO_2 concentration was elevated to 760 ppmv with reduction of soil water content by 35%. In addition we reduced LAI to 2.1.
- Scenario 3: Ambient CO_2 concentration was doubled together with reduction of soil water content by 25%. In addition

we reduced LAI to 2.3 and increased hourly temperature by 1.3°C.

- Scenario 4: Ambient CO_2 concentration was elevated to 760 ppmv with reduction of soil water content by 35%. We reduced LAI to 2.0 and rose the temperature by 2°C.
- Scenario 5: Ambient CO_2 concentration was elevated to 760 ppmv with reduction of soil water content by 50%. We reduced LAI to 1.8 and we rose the temperature by 3°C.

3. Results and Discussion

Air temperature has a regulatory role in governing crop temperature and intensity of biochemical processes. For this reason we examined the canopy air space and crop temperatures in the height of the cob level, where the assimilatory surface is the most developed and the physiological processes the most intensive. If there is no soil moisture limitation, transpiration helps in adjusting crop temperature to air temperature, that is, in bringing their values closer to each other.[1]

A slight reduction in leaf surface size together with 25% decline in soil water decreased both the air temperature inside the canopy and the plant temperature at a very moderate level. In case of the 2nd Scenario it was not necessary to wait (the mean of 0.11°C between 1 and 24 hours) for the more moderate temperature change of inside canopy air temperature, as in the first scenario (the mean of 0.36°C between 1 and 24 hours), see Figure 1. In the 2nd Scenario we applied three times higher leaf area reduction than in the 1st Scenario.

Similar results were obtained in the 5th Scenario (the mean of 2.55°C of canopy air temp. between 1 and 24 hours) where changes in meteorological inputs were the highest. In the case of 5th Scenario the size of variation in simulated temperatures did not reach the increment of the input air temperature, the measure of prognoses in warming. In the 3rd and 4th Scenarios the daily mean of air temperature inside the canopy slightly exceeded the additional air temperature increase (input) (the mean values of 1.59°C and 2.07°C between 1 and 24 hours, respectively) and the daily mean crop temperature was also in the rise. During the daytime hours the increment was lower in the

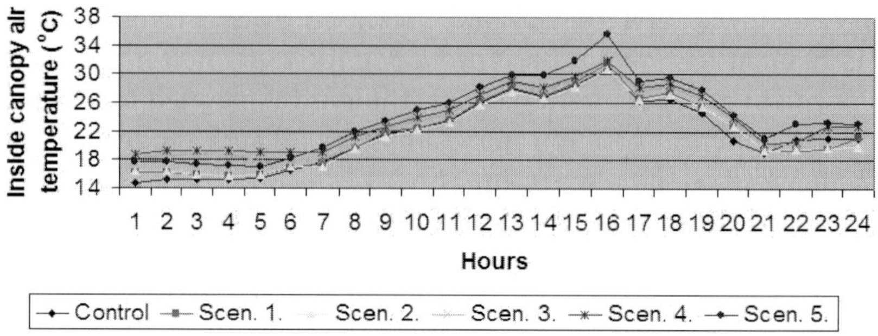

Figure 1. Simulated inside canopy air temperature in corn at varying climate conditions.

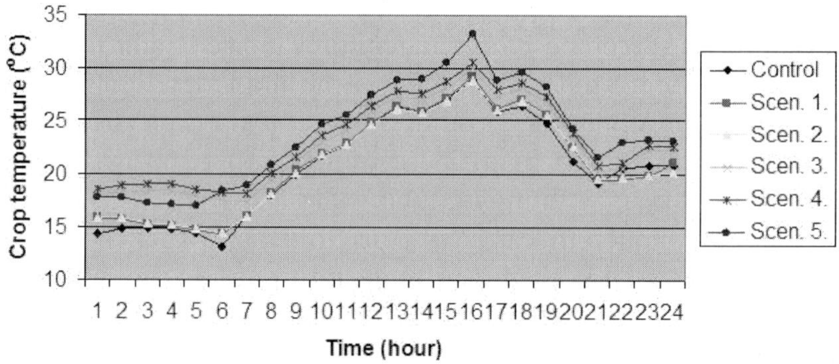

Figure 2. Simulated crop temperature according to different climate scenarios.

case of the air temperature inside the canopy (the mean values of 1.15°C and 1.51°C between 8 and 19 hours, respectively) than the additional input temperature increase, while the rise of crop temperatures was nearly equal to it (the mean values of 1.33°C and 1.89°C between 8 and 19 hours, respectively). During the sunshine hours the canopy itself behaved as a special shelter against the outside warming and the inside canopy temperature increase was more moderate than the rise in the surrounding air temperature, see Figure 2. It is important to mention that the canopy temperature could remain close to the temperature of the surrounding air justifying no existence of heat stress of the plant.

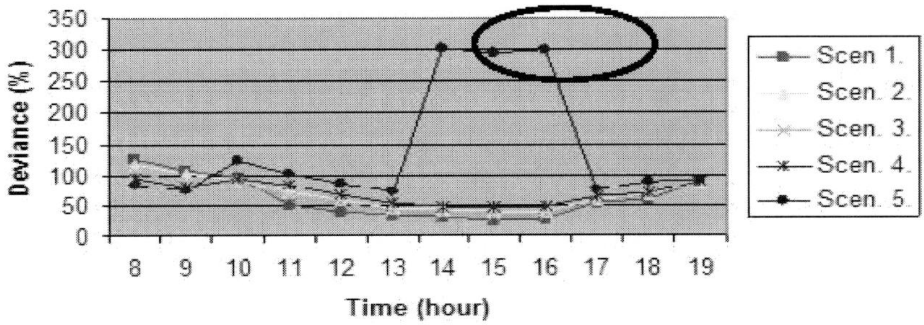

Figure 3. Percentile deviance of stomatal resistance in the different scenarios from the control run.

Some of the scenarios (3rd and 4th) showed just the opposite direction of change by night. They kept more heat in the canopy.

In the 5th Scenario stomata closed between 14 and 16 hours, thus, the cooling possibility of the plant was limited during the hottest period of the day, and the plant could avoid severe water losses, but without cooling (the mean of 2.96°C crop temp. between 1 and 24 hours), see Figure 3. Because of the above-mentioned reason, a higher deviation in crop temperature is to be expected.

Special behaviour of corn inside canopy and plant temperatures might be explained by the size of inputs of LAI and modification in leaf density. As opposed to earlier model simulations, in our study we used "crop" analogies and reduced leaf area index (LAI) in the scenarios simulating stronger levels of global warming. At a reduced transpiration surface, the plant could cool its tissues with less quantity of water. Existing interaction between the plants and their environment suggest that any modification in plant architecture affects not only the microclimate of crops, but their physiological processes (among others transpiration) as well. In every situation the plants strive to maintain the needed equilibrium between physiological properties and the surrounding micro-environment.

References

1. A. Anda, *Commun. Biometry Crop. Sci.* **1**, 90 (2006).
2. J. Bartholy, R. Pongrácz, I. Matyasovszky and V. Schlanger, *AGRO-21 füzetek* **33**, 3 (2004).

3. J. Chen, Mathematical analysis and simulation of crop micrometeorology, Ph.D. Thesis, The Netherlands (1984).
4. J. Goudriaan, Crop micrometeorology: A simulation study, Simulation monographs, Pudoc (1977).
5. S. Solomon, D. Qin, M. Manning, Z. Chen, M. Marquis, K.B. Averyt, M. Tignor and H.L. Miller, *http://www.ipcc.ch/* (2007).
6. D.T. Mihailovic, B. Lalic, I. Arsenic, J. Eitzinger and N. Dusanic, *Ecol. Model.* **147**, 199 (2002).
7. D.T. Mihailovic and J. Eitzinger, *Ecol. Model.* **202**, 465 (2007).
8. J. Mika, *Fizikai Szemle.* **52**, 258 (2002).
9. J. Páll, A. Anda and M. Hunkár, *Acta Geograph., Geologica et Meteorologica Debrecina* **41** (1998).

CHAPTER 25

ESTIMATING WINTER WHEAT YIELD AND PHENOLOGY DYNAMICS USING MET AND ROLL WEATHER GENERATOR

BRANISLAVA LALIĆ AND DRAGUTIN MIHAILOVIĆ

Faculty of Agriculture, University of Novi Sad, Novi Sad, Serbia

MIROSLAV MALEŠEVIĆ

Institute of Field and Vegetable Crops, Novi Sad, Serbia

A four-variable weather data series (1957–1998) for the Rimski Sancevi weather station were calculated/performed. Using statistical parameters of these series, Met and Roll weather generator synthesised weather series for the present and changed climate conditions. For that purposes, climate scenario for Novi Sad (Rimski Sancevi) assimilated from ECHAM4 climate model run at SRES-B1, SRES-A2 and middle emissivity scenario for CO2, have been used. In order to validate the generated weather data an error analysis was performed using generated and observed data for the years from 1990 to 2005. The efficiency of synthetic weather in simulating crop growth was tested using generated and observed weather as input data for SIRIUS crop model. Calculated values of yield and grain mass and dates of growth stages for both series were compared with the observed ones.

1. Introduction

Weather generator (WG) is a stochastic numerical model designed in order to generate daily weather series statistically "identical" to the observed one. However, weather generator is not a predictive tool that can be used in weather forecasting, but it is simply a tool for generating time-series of synthetic weather based on statistical parameters calculated using the observed weather series.

During the first years of its implementation, the WG has been used for agricultural risk assessment estimation and production of weather time-series for unobserved locations through interpolation of weather generator parameters obtained from running the models at neighboring sites.[1] Recently, since outputs from global climate models (GCMs) are of insufficient spatial and temporal resolution, the WG has become a very useful tool in climate-change impact studies.[2–8] In these studies, the generator parameters are modified in accordance with climate change scenarios in order to provide synthetic weather series for the changed climate conditions. The obtained weather series are used as inputs to crop growth models in assessing impact of climate change and elevated CO_2 on crops.[9,10]

In order to identify the most suitable scenario for agro-ecological conditions of Novi Sad region, seven climate scenarios (CSIRO, CGCM2, ECHAM4, GFDL, HadCM3, CCSR/NIES and NCAR-PCM) have been tested in this study. Among six greenhouse gases (GHG) emission scenarios, presented in the Third Assessment Report of IPCCs for the coming century,[11] SRES-B1 and SRES-A2 have been chosen as the "best" and the "worst" scenarios, respectively. Namely, SRES B1 is the lowest emission scenario: it assumes low population growth, low economic growth and rapid technology development. On the other hand, SRES-A2 is a high emission scenario, which assumes low population growth, high economic growth and new and efficient technology. Additionally, "medium/middle emission" scenario (ME) has been used as an intermediate scenario between SRES-B1 and SRES-A2.

In this paper Met and Roll weather generator (MRWG)[12,13] has been run to generate daily series of synthetic weather for Rimski Sancevi (Novi Sad) weather station based on 41 years of observed weather (1957–1998). In order to validate MRWG, synthetic time series were generated assuming different climate scenarios and CO_2 elevations for Novi Sad and compared with the observed ones. Quantitative error analysis was performed comparing relative deviations of monthly temperatures and precipitation sums calculated using the observed and synthesized weather series for the 1990–2005 vegetation period.

The efficiency of synthetic weather in simulating crop yield was tested using the generated and observed weather series as input data for SIRIUS crop model,[14] since this model was already calibrated for NS sorts of small grains.[15,16] The calculated values of yield, grain mass and dates of growth stages during the 2001–2005 vegetation period for both series are compared with the observed ones.

2. Data and Method

Observed data. In order to calibrate MRWG, the daily maximum and minimum temperatures, precipitation and solar radiation at Novi Sad (Rimski Sancevi) weather station were analysed for the 1957–1998 observation period. The period employed for validation of the MRWG was from 1990 to 2005.

The crop data used for crop model calibration were collected during the experiments carried out at the Institute for Field and Vegetable Crops experimental fields (Novi Sad, Serbia) during the 2001–2005 vegetation periods. The selected cultivar was "Anastasia" winter wheat. Among the variety of characteristics, during the five seasons there are recorded data related to time and number of management operations, soil characteristics, phenological dates as well as biomass, grain mass and yield development.

Simulated data. The synthetic series of daily maximum and minimum temperatures, solar radiation and precipitation amounts were generated for Novi Sad (Rimski Sancevi) using MRWG for 1990–2025 periods. Climate scenarios are assimilated from seven climate models (CSIRO (A), CGCM2 (C), ECHAM4 (E), GFDL (G), HadCM3 (H), CCSR/NIES (J), NCAR-PCM (N)) integrated over three different integration periods (2025, 2050 and 2100) and with three CO2 scenarios (SRES-B1 (lo), SRES-A2 (hi), ME (mi)). It means that 63 different scenarios have been used in order to access the best possible scenario for the Novi Sad region. The X-yycc notation was used for each scenario in the text below, where X denotes climate model (A, C, E, G, H, J, N), yy — integration period (25, 50, 00) and cc — emission scenario (hi, lo, mi).

Validation method. In order to conduct qualitative analysis of the generated data a validation procedure should be provided. Taking into account data origin and synthesizing procedure, the comparison of daily data become useless. Namely, daily data are stochastically generated to produce monthly averages equal to the observed ones, but changed in accordance with climate scenario. Otherwise, for each WG run the daily data for particular day will be different but monthly averages will be almost the same. For these reasons and having in mind that for agro-meteorological purposes, the sum of daily values is often more important than a single-day value, monthly sums of the maximum (T_{max}), minimum (T_{min}) and daily temperatures (T_d) as well as the sum of precipitation (H) have been calculated. Since threshold temperature for almost all physiological processes are above 0°C, the temperature sums are calculated with this threshold (T_{d0}). The error analysis was performed by calculating relative error of temperature and precipitation monthly sums during the whole year (Table 1) and for the vegetation period alone (Table 2). On the basis of results presented in Tables 1 and 2, three major conclusions can be made: a) annual relative deviation of the maximum, minimum and daily temperature monthly sums is much higher than for the vegetation period alone; b) climate models run with low emission scenarios integrated till 2025 produce the lowest relative deviation of temperature sums; c) in all cases a relative deviation of synthesized precipitation is remarkable. Therefore, the same reliability should be expected from each of the above-mentioned climate models when run with low emission scenario and till 2025 integration period.

For further calculation in this study, the climate model from the Max-Planck Institute for Meteorology (ECHAM4)[17] has been used. Since the difference between relative deviations of temperature daily sums for "hi" and "lo" scenarios are only 2.6%, this model can be also used for CO_2 impact assessment studies in Vojvodina region.

Table 1. Annual relative error of the maximum (rT_{max}), minimum (rT_{min}), daily temperatures (rT_{d0}) and precipitation (rH) monthly sums averaged over 1990–2005 validation periods (continued on the next page)

Scenario	rT_{max} (%)	rT_{min} (%)	rT_{d0} (%)	rH (%)
A-00hi	91.4	426.2	324.8	260.9
A-25hi	55.6	156.5	142	383.9
A-50hi	64.4	211.8	188.2	324.9
C-00hi	86.1	196.1	219.2	391
C-25hi	54.7	133	127.4	417.6
C-50hi	62.3	148.4	149.6	405.8
E-00hi	111.4	360.8	311	452.1
E-25hi	59.8	149	153.7	438.7
E-50hi	72.5	184.8	200.9	435.1
G-00h	47.5	141.7	107.9	493.5
G-25hi	47.2	139.5	91.9	449
G-50hi	47.5	139.2	118.4	491.6
H-00hi	113.5	322.2	318.5	366.3
H-25hi	59.8	142.5	150.2	419.2
H-50hi	72.6	165.1	176.8	376.5
J-00hi	137.3	499.5	431	230.7
J-25hi	64.4	151.7	170.2	385.7
J-50hi	82.3	235.4	228.1	349.6
N-25hi	56.4	135.3	145.8	269.1
N-50hi	65	155	150.7	253.6
A-00mi	64.4	211.8	191.2	345.5
A-25mi	52.6	146	128.6	499.4
A-50mi	58	164.3	153.7	363.5
C-00mi	62.4	146.1	164.3	462
C-25mi	52.1	136.1	120.7	420
C-50mi	56.6	137.8	134.3	415.6
E-00mi	72.6	184.8	180.8	442
E-25mi	54.9	141.7	138.3	437.1
E-50mi	62.2	151.7	144.8	478.9
G-00mi	47.3	140.8	102.7	462.6
G-25mi	47.4	142.6	97.5	387.7
G-50mi	46.9	145.4	92.8	437.7
H-00mi	72.9	167.8	179.1	377.9
H-25mi	54.8	134.9	124.8	388.7
H-50mi	62.2	143.6	146.4	397.6

(*Continued*)

Table 1. (*Continued*)

Scenario	rT_{max} (%)	rT_{min} (%)	rT_{d0} (%)	rH (%)
J-00mi	83.1	236.6	237	379.5
J-25mi	58.4	137.7	134.7	399.9
J-50mi	67.7	165	202	378.2
N-00m	64.4	152.2	150.7	224.8
N-25m	52.6	132.8	128.2	473.9
N-50m	58	139.1	154.8	362.2
A-00lo	55.6	153.2	136.6	383.9
A-25lo	50.4	135.6	122.3	409.6
A-50lo	52.1	140.2	146.8	398.8
C-00lo	54.6	135.3	152.4	402.2
C-25lo	50	136.5	113.7	426
C-50lo	51.8	135.1	110.6	418.2
E-00lo	59.6	145.6	144.5	437.2
E-25lo	51.8	135.1	121.6	427.2
E-50lo	54.2	139.2	127.9	406.6
G-00lo	48.1	145.3	115.2	445.2
G-25lo	47.4	145	97.9	438.1
G-50lo	47.5	144.7	105.1	443
H-00lo	59.4	143.5	149.6	398.7
H-25lo	52.2	136	109.3	419.8
H-50lo	54.9	139.8	134.4	405.5
J-00lo	64.1	150.3	159.6	356.4
J-25lo	53.8	137.9	112.7	397.9
J-50lo	58.4	138.4	161.1	371.3
N-00lo	56.4	136.6	143.5	317.9
N-25lo	51	131.3	115.7	395.2
N-50lo	52.9	129.7	113.8	393.3

Temperature sums obtained using the observed and generated daily temperatures above 0°C during the vegetation period, see Figure 1, show significant variability over time with strong tendency of the observed curve to "follow" ME or SRES-B1 scenario. Therefore, variability more than increase can be assumed as a vegetation period temperature characteristic in the future. On the other hand, all scenarios predict significantly smaller amount of precipitation during the vegetation period for the next eighteen years.

Table 2. Relative error of the maximum (rT_{max}), minimum (rT_{min}), daily temperatures (rT_{d0}) and precipitation (rH) monthly sums during the vegetation period and averaged over 1990–2005 validation periods (continued on the next page)

Scenario	rT_{max} (%)	rT_{min} (%)	rT_{d0} (%)	rH (%)
A-00hi	30.2	45.3	33.1	228.5
A-25hi	12.1	17	12.4	471.4
A-50hi	16.2	23.6	16.7	360.7
C-00hi	22.5	33.8	24.9	510
C-25hi	11.1	15.9	11.4	542.9
C-50hi	13.5	19.3	13.8	526.2
E-00hi	31.8	65.1	41.1	585.4
E-25hi	11.9	20.2	13.1	572.1
E-50hi	16.4	31.9	19.8	571.8
G-00hi	10.1	15.4	10.7	631.3
G-25hi	10.1	15.4	10.7	583.8
G-50hi	10.2	15.4	10.8	655.7
H-00hi	38.3	61.1	44.3	373.7
H-25hi	13.3	20	13.9	524.1
H-50hi	18.8	30.1	21.2	428.1
J-00hi	32.6	59.3	39.9	188.7
J-25hi	12.1	19.5	13.2	478.4
J-50hi	16.4	29.9	19.3	411.8
N-25hi	11.5	16.9	12.1	258
N-50hi	14.2	22.3	15.4	240.1
A-00mi	16.2	23.7	16.7	396.9
A-25mi	10.9	15.7	11.3	702.6
A-50mi	12.9	18	13	430
C-00mi	13.6	19.5	14.1	642.3
C-25mi	10.5	15.5	10.9	546.2
C-50mi	11.5	16.6	11.9	539.8
E-00mi	16.4	32.1	19.9	576
E-25mi	10.9	17	11.6	571.9
E-50mi	12.5	22.2	14.1	653.8
G-00mi	10	15.5	10.6	597.7
G-25mi	10.2	15.5	10.8	468.3
G-50mi	10.1	15.4	10.7	558.5
H-00mi	19	30.4	21.4	428.4
H-25mi	11.6	16.7	12	466.6
H-50mi	14.1	21.5	14.9	477.6

(*Continued*)

Table 2. (*Continued*)

Scenario	rT_{max} (%)	rT_{min} (%)	rT_{d0} (%)	rH (%)
J-00mi	16.4	29.8	19.5	470.6
J-25mi	11	16.8	11.7	506.1
J-50mi	12.7	21.1	13.9	464.4
N-00mi	14.2	22.2	15.4	184.8
N-25mi	10.7	15.8	11.2	661.1
N-50mi	11.9	17.7	12.6	448.4
A-00lo	12.2	16.9	12.3	472.5
A-25lo	10.4	15.4	10.8	523.5
A-50lo	10.9	15.8	11.3	496.4
C-00lo	11	15.9	11.4	516.1
C-25lo	10.3	15.1	10.6	555.1
C-50lo	10.5	15.3	10.8	542.6
E-00lo	11.9	20.2	13	570.9
E-25lo	10.2	15.6	10.8	552.5
E-50lo	10.8	16.7	11.5	513
G-00lo	10.1	15.4	10.7	579.7
G-25lo	10.1	15.3	10.7	571.5
G-50lo	10.2	15.3	10.7	576.5
H-00lo	13.3	20.1	13.9	489.2
H-25lo	10.5	15.8	11	533
H-50lo	11.4	16.5	11.8	517.2
J-00lo	12	19.4	13.1	420.9
J-25lo	10.5	15.7	11	498.5
J-50lo	10.9	16.6	11.6	447.1
N-00lo	11.5	16.8	12.1	356
N-25lo	10.4	15.2	10.8	500.2
N-50lo	10.7	15.8	11.2	502

3. Results and Discussion

On the basis of all the above presented results, ECHAM4 model with SRES-B1 scenario for CO_2 emission has been used with MWRG to generate future daily weather data. Simulated values of growth stage appearance obtained using the observed and synthesised weather are compared with the observed ones, see Figure 3. For all phenological phases, the difference between the growth stage appearance obtained

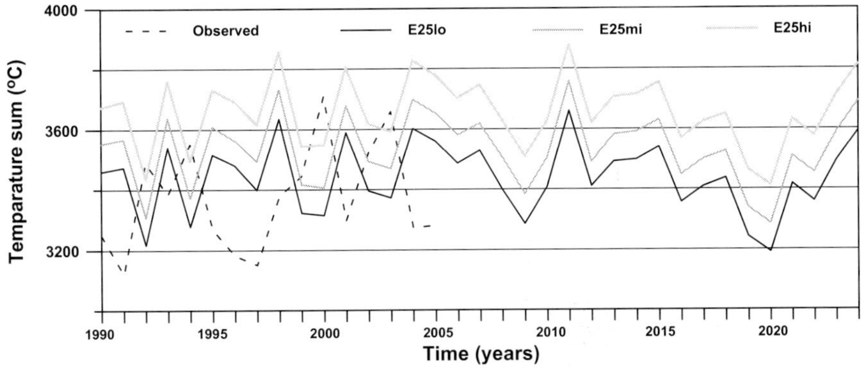

Figure 1. Sum of daily temperatures above 0°C (T_{d0}) during the vegetation period calculated using the observed and generated data.

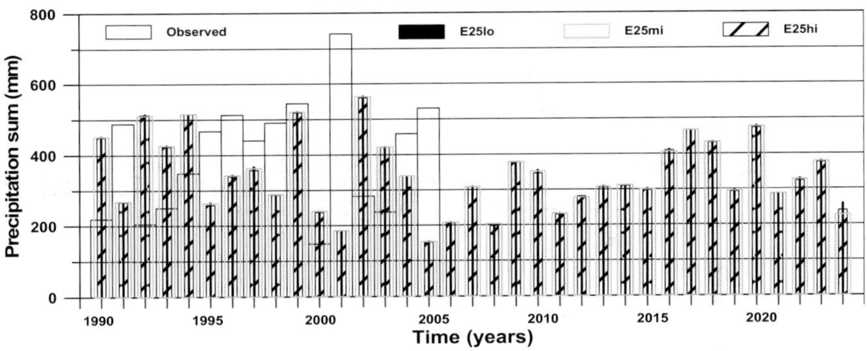

Figure 2. Sum of precipitation (H) during the vegetation period calculated using the observed and generated data.

using the observed and synthesised weather data is non-significant. The average difference between the observed and calculated date of appearance for all phenological phases is 7.7 days for the observed weather and 8.2 days for the generated data. Reliability of synthesised weather in calculating maturity day, i.e. date of harvesting is of special significance since it can strongly affect planning and timing of field management operations.

The results of crop yield simulation, see Figure 4, indicate that deviation of the calculated values from the observed ones is 7% for the

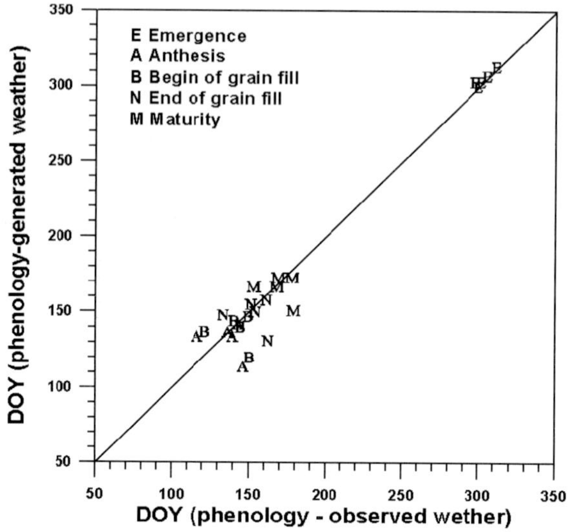

Figure 3. Comparison of growth stage appearance obtained using the observed and synthesised weather data.

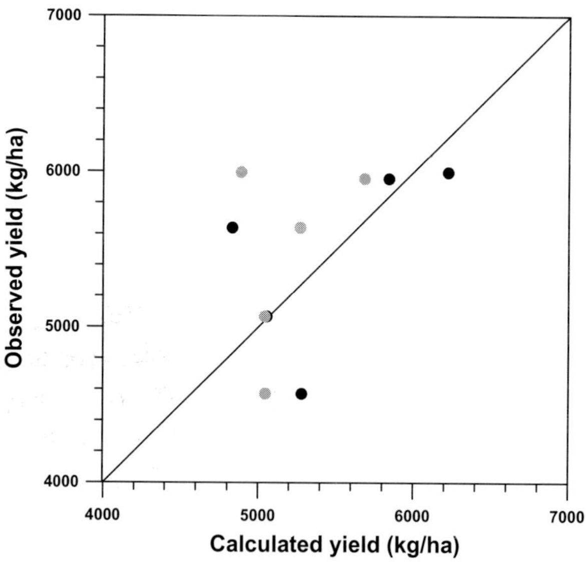

Figure 4. Observed vs. calculated winter wheat yield obtained using the observed (black) and synthesised (grey) weather data.

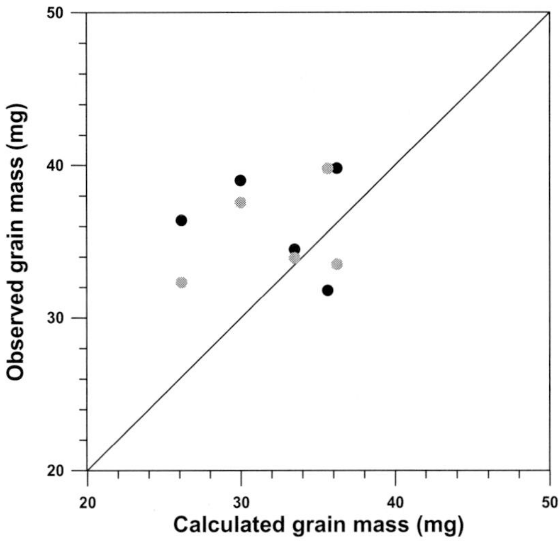

Figure 5. Observed vs. calculated winter wheat grain mass obtained using the observed (black) and synthesised (grey) weather data.

observed weather and 12% for the generated weather. In case of grain mass calculation the obtained deviation is 23% and 18% for the observed and generated weather, respectively see Figure 5.

Although the obtained results are quite satisfactory, further improvement should be made using longer crop data series and introducing new cultivars and varieties.

4. Conclusion

The main goal of this study is to calibrate MRWG and reconsider the possibility to use generated data as input weather data for Sirius crop model in agro-ecological conditions of Novi Sad region. The results obtained in growth stage, yield and grain mass simulations are very promising. Namely, there is neither significant nor systematic deviation between the results obtained using the observed and synthesised weather data. It implies that MRWG can be used as an efficient tool in future climate change impact assessment studies related to wheat production in Vojvodina province.

Acknowledgements

This study was supported by the Serbian Ministry of Science under grant TR6949B. Authors greatly acknowledge Dr Mikhail Semenov and Dr Martin Dubrovsky for their patient assisstance and useful suggestions during the first steps in using weather generators and crop models.

References

1. M.A. Semenov and E.M. Barrow, *Climatic Change* **35**, 397 (1997).
2. M.A. Semenov and J.R. Porter, *Agric. For. Meteorol.* **73**, 265 (1995).
3. S.J. Riha, D.S. Wilks and P. Simoens, *Climatic Change* **32**, 293 (1996).
4. L.O. Mearns, C. Rosenzweig and R. Goldberg, *Climatic Change* **35**, 367 (1997).
5. M.A. Semenov, E.M. Barrow, *Climatic Change* **35**, 397 (1997).
6. M. Dubrovsky, Z. Zalud and M. Stastna, *Climatic Change* **46**, 447 (2000).
7. F.N. Tubiello, M. Donatelli, C. Rosenzweig and C.O. Stockle, *Eur. J. Agron.* **13**, 179 (2000).
8. M. Trnka, M. Dubrovsky and Z. Zalud, *Climatic Change* **64**, 227 (2004).
9. Z. Zalud and M. Dubrovsky, *Theor. Appl. Climatol.* **72**, 85 (2002).
10. S. Bencze, O. Veisz and Z. Bedo, *Cereal. Res. Commun.* 32, 75 (2004).
11. R.T. Watson and the Core Writing Team, IPCC 3rd Assessment Report, *Climate Change 2001*, IPCC, Geneva, Switzerland, 184 (2001).
12. M. Dubrovsky, *Meteorologicke Zpravy.* **49**, 97 (1996).
13. M. Dubrovsky, *Environmetrics.* **8**, 409 (1997).
14. P.D. Jamieson, M.A. Semenov, I.R. Brooking and G.S. Francis, *Europ. J. Agronomy.* **8**. 161 (1998).
15. B. Lalic, L. Pankovic and D.T. Mihailovic, Calibration of SIRIUS model for different NS-sorts of small grains, *Abstracts of 6th Annual Meeting of the European Meteorological Society (EMS)*, Ljubljana, Slovenia (2006).
16. B. Lalic, L. Pankovic, D.T. Mihailovic, M. Malešević and I. Arsenić, *A Periodical of Scientific Research on Field and Vegetable Crops.* **44**, 317 (2007).
17. E. Roeckner and Coauthors, *Tech. Rep.* **218**, 90 (1996).

CHAPTER 26

A REVIEW OF RISK METHODS IN CLIMATE CHANGE IMPACT RESEARCH IN HUNGARY

MÁRTA LADÁNYI AND ÉVA ERDÉLYI

Faculty of Mathematics and Informatics,
Corvinus University of Budapest, Budapest, Hungary

The recent weather anomalies (drought, frost, storm, wind, flood, hail etc.) in Hungary highlighted the vulnerability of agricultural production to climate variability. In this paper we focus on some methods which can be applied to elicit the risk caused by climate change. We introduce some data-management methods and represent some efficiency criteria as well as stochastic efficiency methods we used successfully to detect risk increase. We show some examples of how we combined risk methods with weather indicators in risk management. Using historical data and GCM scenarios we point out some reasons for risk increase in the past and forecast the potential main points of future risks.

1. Introduction

While climate change (CC) causes the climate in Hungary tends to be more and more similar to Mediterranean, extreme weather events are increasing in both magnitude and frequency. Therefore, the risk is of particular meaning as it has been considerably increased in the last few decades. Global CC models, however, project further serious changes in the next 50–100 years including the increase of the mean temperature, increase of spring and autumn precipitation, decrease of summer-winter precipitation, increase of frequency and severity of anomalies, increase of length of extreme periods, increase of uncertainty caused by variability and delayed feedback, increase of CO_2 concentration in the atmosphere, shifted seasons, etc. These long-term changes in climate have numerous potential impacts on production and quality. Climatic risk should be managed by co-ordinated adaptation strategies, preceded by prudent risk analysis.

2. Risk Concepts and Methods

First, we should clarify what is meant by the terms sensitivity, adaptivity and vulnerability. According to the IPCC report[2,6] sensitivity should be defined as how a system responds to given changes in climate. Just think of how sensitive the agricultural production (crop, fruit, vegetable and vine) is to hail storms, frosts, the timing of rainfall (drought or heavy rainfall in growing season, rain at harvest) and warming climate (high maximum and average daily temperature, high minimum night temperature in growing season, etc). Adaptability of a system to CC is shown by the degree of how it can be adjusted (spontaneously or premeditatedly) to changes by sufficient management. Vulnerability of a system can be measured by the unavoidable harm or damage, which depends not only on the system's sensitivity but also on its adaptivity to the new climate conditions. Sensitivity, adaptivity and vulnerability, together with the uncertainty of the future conditions define how risky a particular system is.

2.1. *Data and Data Management Methods*

We used historical weather data with the minimum, maximum, average, precipitation and radiation daily data as well as production data for several Hungarian regions for corn, wheat and vine. For the scenarios approach we applied the GCMs of GFDL (Geophysical Fluid Dynamics Laboratory, USA) which were downscaled to Hungary and refer to about 2030 and 2060.

During our work we were faced with the data-related problems — the available historical data were either too short, not detailed enough or not relevant. To avoid the problem of data absence or shortage we used some very effective data-management methods known in specialised literature, such as:

- Phillips-method for making the data temporary and spatially relevant,[13]
- Bayes- and Monte-Carlo-methods for making the data abundant,[5]
- Delphi-method for the synthesis of experts' opinion[11]
- Smoothing methods for distribution determination
- Triangle-method for probability assessment,[5]
- Rank correlation or Fackler stochastic dependency methods to determine the dependency of the risk variables.[3]

2.2. *Risk Analysis Concepts and Methods*

2.2.1. *Certainty equivalent, utility function and risk aversion*

Certainty equivalent (*CE*) is a certain payoff value which is equivalent, for the decision maker (*DM*), to a particular payoff distribution. If the *DM* can determine his or her *CE* for the payoff distribution of each alternative in a decision problem, then the optimum is the one with the highest *CE*. Therefore, by having *CE* we can not only rank risky alternatives but also estimate the cost and/or the premium of risk. The *CE*, i.e. the minimum selling price for a payoff distribution, depends on the *DM*'s personal attitude toward risk. A *DM* may be risk preferring, risk neutral, or risk avoiding, in most cases, however, the latter one.

A utility function, *U*, can be used to represent a *DM*'s attitude toward risk. Nevertheless, measuring risk preferences can directly lead to unstable results. Besides possible interview-errors, the main problem is that the estimations based on historical data with more or less relevance are mostly different from the subjective estimations of the experts who are interested in the decision. In some cases the *DM* can not even reveal his own attitude to risk. Therefore, we should fix some properties that are expected to be held by a utility function *U* and define a general utility function.

Assuming that we express utility as a function of total return t, we require that utility function should be invariant to positive linear transformation. Absolute risk aversion r_a can now be defined as:

$$r_a : t \mapsto -\frac{U^{(2)}(t)}{U^{(1)}(t)} \tag{1}$$

from which we can conclude that r_a decreases with increasing t. Quite typical and the most widely applied utility function is the so-called negative exponential utility function:

$$U : t \mapsto U(t) = 1 - \exp(-ct) \tag{2}$$

The most important properties of this function are that its absolute risk aversion r_a is constant while its relative risk aversion $r_r = t\, r_a(t)$ is linearly dependent of total return t: $r_a(t) = c$, $r_r(t) = ct$. Taking the Taylor series expansion of the utility function U by a given region around t and having $E(t) = E$, we can approximate the values of U such as:

$$U(t) \approx \frac{U(E) + U^{(1)}(E)(t-E) + U^{(2)}(E)(t-E)^2}{2} \tag{3}$$

As normal distribution is completely specified by the mean E and variance V, in the case when the distribution of t is normal, the so-called Freund-approximation for CE can be used as[4]:

$$CE = E - 0.5 r_a(t) V \tag{4}$$

2.2.2. *Efficiency criterion (E-V-efficiency)*

The E-V efficiency criterion is used when allocation of resources across an array of risky alternatives is needed. We intend to find the set in which the expected total return is maximised for different levels of variance of the total return. If $E_1 \geq E_2$ and $V_1 \leq V_2$, then the alternative belonging to (E_1, V_1) is more preferable. The E-V efficient set includes the portfolio that maximises the expected utility if the *DM*'s utility function is quadratic or the total return is normally distributed.[1,5] However, the method is quite robust to violation of normality. Nevertheless, the *E-V* efficiency criteria leaves usually some alternatives unranked.[10,12]

2.2.3. *Stochastic dominance (1st and 2nd degree)*

Stochastic dominance involves comparing points on more entire distributions. Namely, if $F_1(x) \leq F_2(x)$ (1st degree) or $\int_{-\infty}^{x} F_1(t)dt \leq \int_{-\infty}^{x} F_2(t)dt$ (2nd degree) then the first alternative is better than the second one. The 2nd degree stochastic dominance is applicable when the *DM* is risk averse and has more discriminatory power than the 1st degree stochastic dominance. Its disadvantage is that the assumption of risk aversion may not hold and often happens that after a tiresome computation we cannot rank the distribution functions, and thus, the set of alternatives remains too large.[1,14]

3. Results

The historical production data were corrected with the assistance of Phillips-method in order to make them comparable.[13] Then three efficiency criteria were considered, namely the *E-V*-efficiency, the stochastic dominance based on the subjective distribution functions and the criterion based on the utility function. However, the results have not fulfilled all our expectations. The risk increase was finally proved by a recently simplified variant of the general stochastic dominance criterion.

As the exact personal risk aversion was not known, we used the widely applied negative exponential utility function. The utility function was calculated depending on the degree of risk aversion r_a as: $U(x, r_a) = \int U(t, r_a) \cdot f(t)dt$. The certainty equivalent *CE* was defined as: $CE(x, r_a) = U^{-1}(x, r_a)$.

Under the assumption of a negative exponential utility function the utility function U can be approximated by:

$$U(x, r_a) = \sum_i (F_{i+1} - F_i)\left[1 - \frac{(\exp(-r_a x_i) - \exp(-r_a x_{i+1}))}{r_a(x_{i+1} - x_i)}\right] \tag{5}$$

Certainty equivalent CE can be obtained as the inverse U by the formula:

$$CE = \frac{-\ln[1 - U(x, r_a)]}{r_a} \tag{6}$$

If we represent the graph of certainty equivalent CE depending on the absolute risk aversion r_a, the highest curve indicates the less risky time series.

Applying the above listed data management and risk methods we obtained the following results:

1. In the graphs of corn and wheat yield we recognised that beside the yield loss caused by the Hungarian political situation at the end of the eighties, the deviation of the yield started to become greater already at the beginning of the eighties. There was a heavy yield loss in 1990 caused by several extreme weather events. We investigated the risk within three twenty-year periods (1951–1970, 1961–1980 and 1971–1990). With the help of the above graphic representation we proved that the risk in corn and wheat production increased between 1951 and 1990 in all the examined Hungarian counties, independently of the rate of risk aversion. In some regions the rate of increase was even quicker.[7]
2. We proved that the risk of vine production increased between 1964 and 2000, independently of the rate of risk aversion. In some regions the rate of increase was even quicker. Though some of the examined regions are quite different from each other with respect to their climate, terrain as well as vine production structure, the fact of risk increase can undoubtedly be proven for all of them.[8] In order to see the trends, the time interval 1964–2000

was split into four parts: 1964–1976, 1970–1982, 1976–1988 and 1988–2000 and the risk in vine production was elicited for each of the short intervals:

TS1: 1964–1976 no symptoms of CC;
TS2: 1970–1982 extreme high/low production;
TS3: 1976–1988 yield loss occurs frequently;
TS4: 1988–2000 serious extreme events, warming.

3. Based on the international references and Hungarian experts' opinion we have drawn up a list of the most important, plant specific weather indicators (corn, wheat, and vine). We do not state that production (quality and quantity) is determined by these factors alone and that there are no other important indicators. We just think that agricultural production depends very much on these factors and that a very strong, mainly very obvious connection can be detected between the years with high/low amount and/or quality of production and the values of weather indicators in the historical data. Therefore, the more we learn about these indicators, the more we can conclude about the production as well as the expected risk.[15]
4. In order to collect, organise, manage and search databases for CC research in an attractive and friendly way, a special data management system was needed. Indicator-search software has been developed by K. Szenteleki.[16] The system allows researchers from various disciplines to set up special databases for specific studies and has the capacity to filter and aggregate data from different perspectives. Applying the so-called Climate Profile Indicator Module we can create weather indicators by defining lower and upper boundary conditions regarding daily and monthly data. It is also possible to combine temperature (minimum, average, maximum) as well as precipitation and radiation data of any time period. In the case of daily data, the system of conditions can be set up by day, while for making parameters for longer time periods (weeks, for example), linear interpolation can be applied. The module makes it possible e.g. to define conditions for plant production demands. The software can survey and evaluate past and

predicted future scenarios to monitor temperature and precipitation characteristics and helps decide whether the examined variables (of the examined time series) indicate sufficient or not sufficient conditions, according to the profile indicators.

5. With the assistance of the above-mentioned software we reveal how the appearance of extreme values of the outlined indicators has been changing with time. There is no doubt that the anomalies of the indicators are becoming more and more frequent. Moreover, based on the most widely accepted GCM scenarios we have pointed out that such kind of extremes are expected to become more frequent in the next 30–50 years.[9]

With the extreme event approach, we intend to make the detailed analyses in future on how and to what extent this change is going to proceed. Consequently, we are going to show the connection between the occurrences of extreme events in weather and production fluctuations in quality and quantity. Using historical data and GCM scenarios we will be able to detect the most probable reasons of risk increase in the past as accurately as possible and forecast the potential main points of future risk. This approach will enable us to respond to CC with a National CC Adaptation Programme which is still under construction.

Acknowledgements

Our work was supported by NKFP–B3–2006–0014, OTKA T042583 tender and the Jedlik Ányos NKFP6–00079/2005 programme.

References

1. J.R. Anderson, J.L. Dillonand and J.B. Hardaker, *Agricultural Decision Analysis* (Iowa State University Press, 1977).
2. Allen Consulting Group: Climate Change, Risk and Vulnerability. Promoting an efficient adaptation response in Australia, Final Report (2005).
3. P.L. Fackler, *Am. J. Agr. Econ.* **73**, 1091 (1991).

4. R.J. Freund, *Econometria*. **24**, 161 (1956).
5. J.B. Hardaker, R.B.M. Huirne, J.R. Anderson and G. Lien, *Coping with Risk in Agriculture* (CABI Publishing, 2004).
6. Working Group II, Impacts, Adaptation and Vulnerability, IPCC Report on Climate Change (2001).
7. M. Ladányi and É. Erdélyi, The increase of risk in maize production detected by a new stochastic efficiency method, *Agrárinformatika 2005*, Debrecen (2005).
8. M. Ladányi, É. Erdélyi and K. Szenteleki, The increase of the risk in Hungarian vine production due to climate change, *EFITA Conference*, Glasgow, Scotland (2007).
9. M. Ladányi, É. Erdélyiand and K. Szenteleki, Weather indicators and their tendencies caused by climate change. *OIV Conference*, Budapest, Hungary (2007).
10. H. Levy and H.M. Markowitz, *Am. Econ. Rev.* **69**, 308 (1979).
11. H.A. Linstone and M. Turoff (Eds.), The Delphi method: Techniques and applications, New Jersey Institute of Technology (2002).
12. H.M. Markowitz, *Mean-Variance Analysis in Portfolio Choice and Capital Markets* (Basil Blackwell, 1990).
13. J.B. Phillips, *Statistical Methods in Systems Analysis* 34 (Wiley, 1971).
14. J.W. Richardson, K. Schumann and P. Feldman, *Agr. Econ-Blackwell*, (2001).
15. K. Szenteleki, E.P. Botos, A. Szabó, Cs. Horváth, L. Martinovich and Z. Katona, Definition of the ecological facilities, ecological indicators and quality of products int he Hungarian vine and wine sector using updated GIS, *EFITA Conference*, Glasgow, Scotland (2007).
16. K. Szenteleki, M. Ladányi, É. Szabó, L. Horváth, L. Hufnagel and A. Révész, A climate research database management software. *EFITA Conference*, Glasgow, Scotland (2007).

4. R. [illegible], [illegible] 24, 301 (1978).
5. [illegible], [illegible] and G. [illegible], [illegible] 2006.
6. [illegible] (2007).
7. [illegible] and [illegible] [illegible] change production [illegible] 2005.
8. [illegible]

CHAPTER 27

ECOLOGY AND RISK MANAGEMENT OF HAZARDS IN PHARMACEUTICAL INDUSTRY

JASMINA KANAZIR
AND MIRJANA VOJINOVIĆ-MILORADOV

ACIMSI — Centre for Environmental Engineering, University of Novi Sad, Novi Sad, Serbia

DORDE BAŠIĆ

Institute of Energy, Process and Environmental Engineering, University of Novi Sad, Novi Sad, Serbia

The objective of risk management is to prevent incidents due to the operation of facilities which handle hazardous materials. There is a wide range of available models for risk analysis which help to understand the risk of a hazardous facility. Among many of them, ALOHA (Areal Locations of Hazardous Atmospheres) computer programme, which uses site-specific information provided by its user and physical property data from its extensive chemical library, was used to predict atmospheric dispersion of a hazardous gas cloud after an accidental chemical release. The paper compares the predicted hazard zones as a function of wind speed following the accidental release scenario. The ALOHA atmospheric dispersion model was run in pharmaceutical industry for direct release of acetone under different weather conditions. The movement of pollutants is governed by motions in the atmosphere. Some of the transport phenomena determine the path that the airborne contamination will follow. By defining the width of vulnerable and dangerous zones, potentially vulnerable objects can be identified and the response procedure to accidental situation can be defined.

Similar studies were conducted in different branches of industry in Serbia.

1. Introduction

Over the past few decades, there has been an increasing concern that human actions and natural catastrophes have been adversely impacting the environment and posing serious ecological and health hazards.[1] Ecological risk assessment is a flexible process for organizing and analyzing data, assumptions and uncertainties to evaluate the probability of adverse ecological effects that are occurring or may occur as a result of exposure to one or more stressors related to human activities.[2]

There is no such thing as zero risk because no matter what precautionary steps are taken, there is always some chance of an accidental release of a hazardous substance and a chance that someone will be adversely affected. The objective of environmental risk management is to prevent or reduce the illness, injury or loss of life due to the operation of facilities which handle hazardous materials.

Risk includes a consideration of both the likelihood and severity of an event. Parameters that influence the risk level include the hazards of the material, the quantity of the material involved, the containment system, the type of stress applied to the container, the proximity of exposures, and the level of available response resources, including the level of training and the experience of the responders.[3] There are hazardous materials in every community. While most releases are minor, there is always a high potential for a significant release which could have direct effect on public health and environment. The risks associated with a chemical release in a community are not confined to city or country boundaries. The "worst case" release could affect an area across any border.[4] Major chemical accidents cannot be predicted solely through regulatory requirements. Rather, understanding the fundamental root causes, widely disseminating the lessons learned, integrating these lessons into safe operations and good practice in environmental management is also required.[5]

2. Understanding the Hazard

An ecological risk management exists within a larger context which encompasses risk assessment and risk communication. Risk management initially sets the boundaries of the assessment providing a way to develop, organise and present scientific information so that it is relevant to environmental decisions.[2]

The first step in managing chemical reactivity hazards is identifying those facility operations and chemicals that represent a potential chemical hazard. Most of the efforts focus on ensuring that the facility is designed and operated safely on a day-to-day basis, using safely designed equipment, preventive maintenance, up-to-date operating procedures and well-trained staff.[6]

When selecting the hazardous events for quantifying risks, it is important to include a range of scenarios, from the worst possible case to more realistic scenarios. Under normal circumstances, the theory of atmospheric dispersion, diffusion and transport is a tool in studying the disposition of hazards in the atmosphere and it is applicable to a variety of release conditions, including unintentional releases of hazards which are potentially harmful for humans, biota and environment in general.

The fundamental reason for modelling atmospheric transport and diffusion is to predict the concentration in the atmosphere of hazardous material released from a source as the material moves from the source to other locations. Hypothetical release at particular points and under different conditions can be used by planners to identify potential zones of hazardous threats and prepare effective responses to these scenarios.[7]

Risk analysis results are always derived from processing of large quantities of information obtained from numerous scientific fields. There is a wide range of available models and the type of a model appropriate for a particular use depends on the scale of the problem, the level of detail available for input and required for output, the background of the intended user and the time available to wait for the model computations to be completed.

ALOHA computer programme has been designed to require input data that are either easily obtained or estimated at the scene of an accident. ALOHA can model the dispersion of a cloud of pollutant gas in the atmosphere and display a diagram that shows threat zones. There are two separate dispersion models in ALOHA: Gaussian and heavy gas, depending on the characteristics of the released hazard. ALOHA can automatically choose whether to predict the dispersion of a chemical as a Gaussian or heavy gas release. ALOHA allows modelling of releases from four types of sources: Direct, Puddle, Tank and Gas Pipeline.[8]

ALOHA software and Gaussian model of estimated concentration was applied in the case of urban area of Kragujevac, which had suffered the greatest devastation during NATO intervention. During a response to hazardous chemical accidents, ALOHA proved to be a good prediction tool.[9]

3. Experimental Work

Urban air pollution is a common problem facing big cities, especially in the developing countries such as Serbia. Because air quality is greatly affected by weather conditions, air quality can be very bad during the days when weather conditions are unfavourable for the transport and diffusion of pollutants. Therefore it is important to consider the weather conditions in the decision-making process.[10]

This paper compares the predicted hazard zones as the function of wind speed following the accidental release scenario on the example of the chosen pharmaceutical plant in the territory of Serbia that has been operating successfully for many years. The plant is set in the close vicinity of the town of Pančevo and poses the possible threat for both the industry and its surrounding in the case of direct release of acetone. Acetone is commonly used in the chosen pharmaceutical industry as a solvent for paints. Acetone quickly evaporates to a gas if released as a liquid and exposure to this hazard can cause some serious symptoms such as headache, fatigue, liver and lung damage, unconsciousness and death.[11] Chemical characteristics of the analysed hazard are given in the Table 1.

Table 1. Chemical characteristics of analysed hazard

	Acetone
Molecular weight, g/mol	58.08
Ambient boiling point °C	55.70
Vapour pressure, atm	0.14
Ambient saturation concentration, %	14.10
TEEL-1, ppm	1000.00
TEEL-2, ppm	8500.00
TEEL-3, ppm	8500.00

Abbreviation TEEL means Temporary Emergency Exposure Limit to maximum concentrations in the air below which is believed all individuals could be exposed without:

- Experiencing other than mild transient adverse health effects or perceiving a clearly defined objectionable odour (TEEL-1),
- Experiencing or developing irreversible or other serious health effects or symptoms that could impair their abilities to take protective action (TEEL-2),
- Experiencing or developing life-threatening health effects (TEEL-3).

The ALOHA atmospheric dispersion model was run for direct release of acetone under different weather conditions. Experimental weather conditions were selected in such a way to cover most scenarios that could conceivably occur under normal circumstances. The plant is situated in the area where fluctuations in wind speed are a common phenomenon, with near gale force winds. At low wind speed the model used heavy gas dispersion algorithm, while at high wind speed a Gaussian model was employed. Table 2 summarises the relevant release parameters.

Direct dispersion of acetone in relation to the wind speed and time of release is given in the figures below. A day with strong, near gale force wind was selected as a simulation period. The influence of ground roughness on the outdoor concentration of released acetone

Table 2. Release scenarios used in the dispersion estimates

Storage temperature	Ambient, 12°C
Time of release	10, 30 and 60 minutes
Wind speed	3 m/s and 15 ms^{-1}
Stability class	D (for completely overcast conditions during day or night)
Relative humidity	32 %
Ground roughness	Urban or forest
Accidental simulation	Outdoor and indoor
Pollutant	Acetone
Daily storage	5600 kg
Amount of pollutant for the specified time of release	560 kg
PEL	1000 ppm

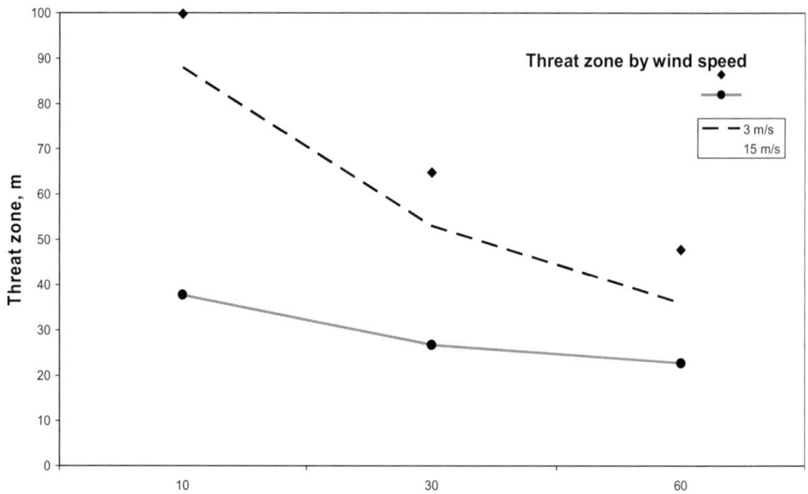

Figure 1. Illustration of the threat zone at different duration of release.

was insignificant. Stability class has a big effect on the prediction of the threat zone size, since the plume cloud will not extend as far downwind as it would under more stable conditions, as in stability class D. As shown in Figure 1, at high wind speed the length of the threat zone is shorter due to sooner mixing of dispersing acetone with the air around it, which is in close connection with the stability class D.

Despite important differences in the atmospheric terms, the behaviour of the threat zone as the function of wind speed for the chosen hazard is remarkably similar, as shown in Figures 2 and 3. Even though the large amount of acetone is released into the atmosphere it is clearly seen that the increasing of vulnerability distance will result in reduction of outdoor concentration of acetone below its permissible level of concern. At higher wind (15 ms^{-1}) the outdoor concentration attains not even the permissible level of concern for acetone which results in lower rates of accident due to better atmospheric dispersion.

It is obvious that under the worst-case scenario, an accidental release may spread over the industrial complex site in a long-term and in the parts of the nearby town in a short-term interval, which means that weather conditions are very important for short-term control of

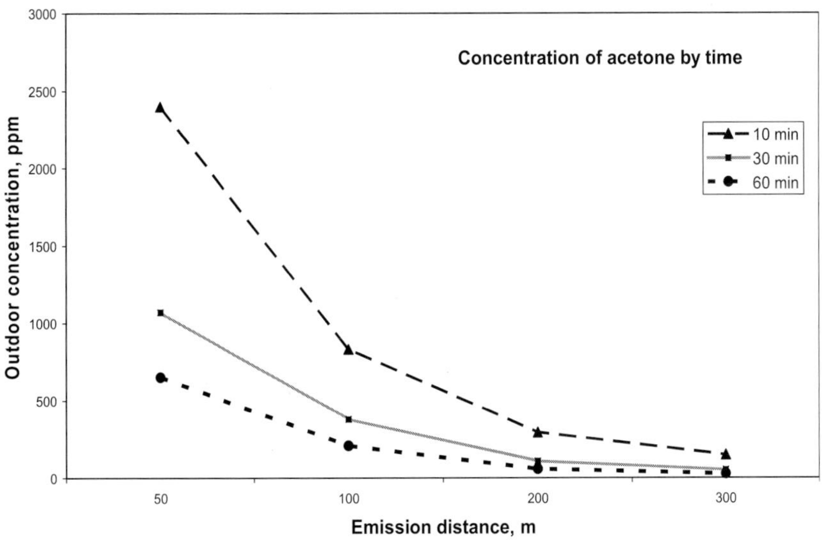

Figure 2. Illustration of the impact of outdoor concentration of acetone at wind speed of 3 m/s.

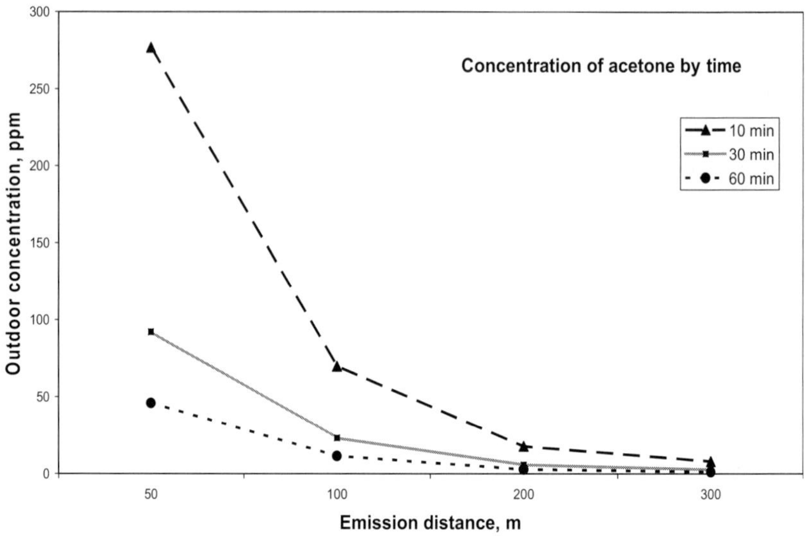

Figure 3. Illustration of the impact of outdoor concentration of acetone at wind speed of 15 m/s.

air pollution. With the increasing distance from the accident source, the outdoor concentration decreases, thus indicating the short-term effect in some parts of the town.

Even though the consequences of the possible accident could be insignificant, it is recommended to stay indoors after the warning alarm. Frequent changes in wind direction represent additional extenuating circumstances, which contribute to better dispersion of pollutants.

4. Conclusion

The fact that acetone is stored in the tanks in the amounts sufficient to meet monthly and even annual consumption, warns, by using ALOHA mathematical prediction, that there is a continuous likelihood of accidental situations which could seriously affect human health and the environment in general.

Critical distance for the low speed wind is from 50 to 100m in a 10-minute release, while in the case of high wind speed outdoor concentration of acetone the permissible level of concern is not reached. Thus, obtained results confirm that acetone has relatively low risk assessment for the chosen industrial complex. Furthermore, it is to be taken into account that the chosen pharmaceutical industry operates in inadequate conditions of transition and former economic isolation. Knowing and understanding the risk of hazardous facility and hazard release are the most important segments of an optimal environmental management. An effective hazard and risk assessment allows the operator to develop an incident action plan and implement strategies and tactics.

References

1. N. Raheja, http://www.gisdevelopment.net (2007).
2. B.K. Hope, *Environ. Int.* **32**, 983 (2006).
3. Hazard and Risk Evaluation, http://www.efilmgroup.com/Eight-Steps-to-Hazardous-Materials-Management.html (2007).
4. Safety through communication and education, http://www.tus-coema.org (2007).
5. EPA, Identifying chemical reactivity hazards: Preliminary screening method, Office of Emergency Prevention, Preparedness and Response, EPA 550-F-04-004 (2004).
6. EPA, Chemical Accident Prevention: Site Security, Office of Emergency Prevention, Preparedness and Response, EPA K-550-F00-002 (2000).
7. M.J. Borysiewicz and M.A. Borysiewicz, Models and techniques for health and environmental hazard assessment and management, Center of Excellence for MANHAZ, Institute of Atomic Energy, Poland (2006).
8. ALOHA User's manual, U.S. Environmental Protection Agency (2006).
9. M. Vojinovic-Miloradov, M. Turk, J. Radonic, Modeling of transport phenomena of PCB congeners through the atmosphere, *The 7th European Meeting on Environmental Chemistry*, Brno, Czech Republic, 22 (2006).
10. F. Liu, J. Zhu and Y. Zhang, *Environ. Mod. Softw.* **22**, 548 (2007).
11. Acetone and dichloromethane fact sheet, http://www.npi.gov.au/database/substance-info/profiles (2006).

CHAPTER 28

UNCERTAINTIES IN THE ESTIMATION OF A SHALLOW LAKE WATER BUDGET

TAMÁS WEIDINGER* AND ÁRPÁD BORDÁS*,†

**Department of Meteorology, Eötvös Loránd University, Budapest, Hungary*

†*University Centre for Meteorology and Environmental Modelling, University of Novi Sad, Novi Sad, Serbia*

SZILVIA SIMON AND JUDIT MÁDLNÉ SZŐNYI

Department of Physical and Applied Geology, Eötvös Loránd University, Budapest, Hungary

The estimation of evaporation and water budget of inland water surfaces are important for better understanding of interrelationship between the surface water bodies and groundwater flow systems. The aim of our study was to determine the water budget of a shallow (40–60 cm deep) lake and to allocate the origin of the high salt content of the lake water using micrometeorological measurements. The subject of our study is Lake Kelemen-szék (Hungary), located in the discharge area of a gravity-driven and an over-pressured hydraulic system. The micrometeorological measuring tower was installed 40 m offshore in the lake. The continuous and simultaneous measurements of global and net radiation, temperature, relative humidity and wind speed gradients, wind direction, precipitation, and determination of water temperature profile started in August 2006. The diurnal and monthly variations of evaporation were determined using three different methods based on micrometeorological measurements: (i) the energy budget, (ii) the Bowen ratio and (iii) the profile method. The results obtained through different methodologies were compared.

1. Introduction

The goal of our study was to estimate evaporation and determine the water budget of a shallow lake with the idea to understand the interrelationship between the lake and groundwater flow systems, as well as to allocate the origin of the high salt content of the lake water. The subject of our study was the shallow (40–60 cm deep), highly saline lake — Lake Kelemen-szék. The lake is located in the Danube – Tisza Interfluve (Hungarian Great Plain) and belongs to the Kiskunság National Park. The location of the lake is presented in Figure 1. The average area of the water surface is ≈ 2.5 km^2.

The fore going studies[1] showed that two driving forces are prevalent in the Danube-Tisza Interfluve. Near the surface, there are gravity-driven flow systems, which are in the deep, over-pressured

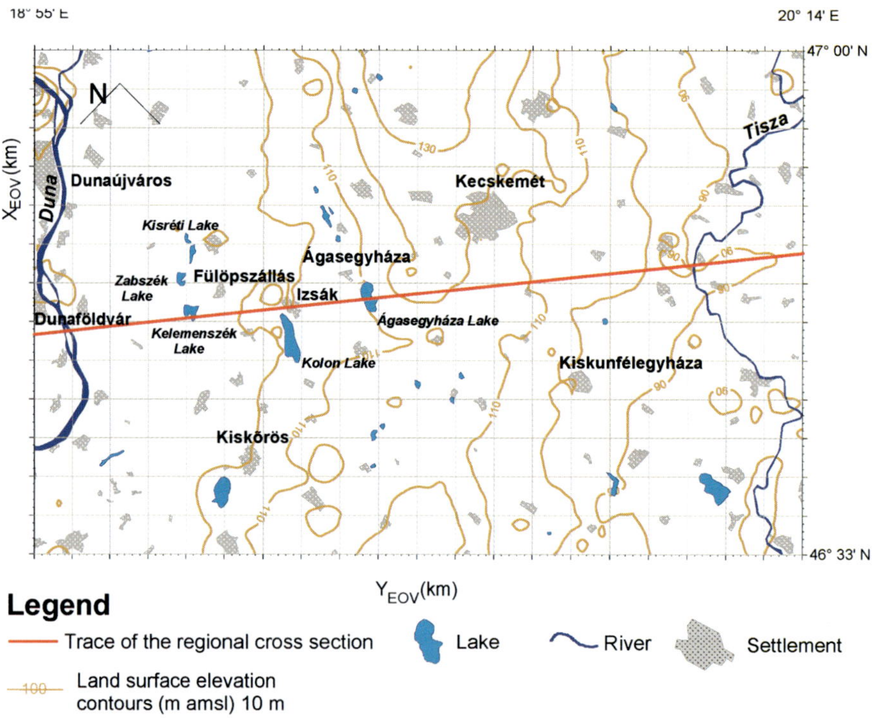

Figure 1. The Danube-Tisza Interfluve and the location of Lake Kelemen-szék.

waters, originating from the preneogene basin fill. Detailed hydraulic and hydrostratigraphic studies demonstrated that the groundwater flow systems reach the surface in the Danube valley, where wetland areas with highly saline lakes can be found. Lake Kelemen-szék is located in the discharge area of a gravity-driven and an over-pressured flow system as well. Our hypothesis was that the high salt content of the lake, and alkaline soils around the lake, originated partly from the highly saline water of the deep over-pressured flow system. This hypothesis refuted the former theory explaining the high salinity only as the result of the intensive evaporation in the examined area. Our objective was to estimate evaporation using micrometeorological measurement data and to verify our hypothesis.

2. Experimental Set-up

The micrometeorological measuring tower was installed 40 m offshore in the lake. Figure 2 shows the shape of Lake Kelemen-Szék and the position of micrometeorological tower. The continuous and simultaneous measurements of global and net radiation (with Schenk pyranometer and Q7 net radiation sensor), temperature, relative humidity and wind speed gradients (using Vaisala HMP 45AC probes, Vaisala anemometer and Young wind monitor at 1 m and 3 m above the water surface), wind direction (at 3 m above the water surface), and determination of water temperature profile started in

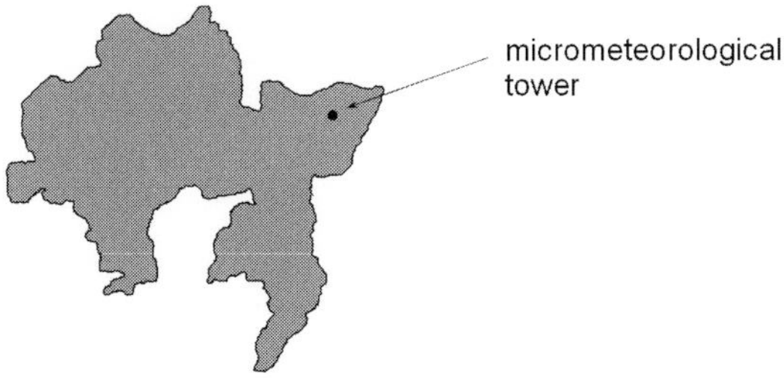

Figure 2. The shape of Lake Kelemen-szék and the position of the micrometeorological tower.

Figure 3. Micrometeorological tower with the equipment: Schenk pyranometer, Q7 net radiation sensor, Vaisala HMP 45AC humidity and temperature probes, Vaisala anemometer, Young wind monitor (on the top of the mast) and CR10 data logger (in the box). A lightning conductor was also installed next to the micrometeorological tower.

August 2006. All data were collected by the CR10 Campbell data logger. The micrometeorological tower and the equipment used are shown in Figure 3. The daily precipitation and water level changes were also measured.

3. Theoretical Background

The diurnal variations of evaporation were determined using micrometeorological measurement data. In this section, three different methods (the energy budget, the Bowen ratio and the profile method) used to calculate evaporation and causes of uncertainties were described.

3.1. *The Energy Budget Method*

The energy budget calculations are based on the energy budget equation:

$$R_n - S = LE + H, \tag{1}$$

where R_n is the net radiation (the sum of the short wave and the long wave radiation budget), S denotes the subsurface heat flux, LE and H are the latent and the sensible heat fluxes, respectively. The difference between the net radiation and the subsurface heat flux (R_n – S) is often called the available energy. We can calculate the subsurface heat flux, the heat entering the water, using water temperature profile measurements as:

$$S = c_w \frac{(\Delta T_a \tilde{m}_a + \Delta T_b \tilde{m}_b)}{\Delta t}, \tag{2}$$

where c_w is the specific heat capacity for water, ΔT_a and ΔT_b are the water temperature changes for the upper (from the water surface to –4 cm) and the lower water level (from –4 cm to –30 cm), $\tilde{m}_a$ and $\tilde{m}_b$ are the masses of upper and lower water layers, and Δt is the time step. The daily average of the sensible heat flux (H) above the water is often lower than the uncertainty in the energy budget calculations. Because of this, in the first approximation we can neglect the sensible heat flux (H = 0) and calculate the latent heat flux as:

$$LE_{eb} = R_n - S. \tag{3}$$

3.2. *The Bowen Ratio Method*

The Bowen ratio method is widely used in determining evaporation. It is popular because it requires only the measurements of temperature and humidity at two levels, net radiation and subsurface heat flux. The Bowen ratio is defined as the ratio of the sensible (H) and the latent heat flux (LE):

$$Bo = \frac{H}{LE}. \tag{4}$$

We can calculate the sensible and latent heat fluxes as:

$$H = -K_H \rho_m c_p \frac{\Delta T}{\Delta z} \quad \text{and} \quad LE = -K_E \rho_m L \frac{\Delta q}{\Delta z}, \tag{5}$$

where K_H and K_E are the turbulent diffusion coefficients for the sensible and the latent heat respectively, ρ_m is the air density, c_p is the specific heat capacity at constant pressure, L is the latent heat of vaporization, Δq and ΔT are the differences of the specific humidity and temperature measured at two different levels, and ΔZ denotes the height difference between levels where the specific humidity and temperature are estimated. In the first approximation it was assumed that $K_E = K_H$ and the Bowen ratio was written as:

$$Bo = \text{const.}\frac{\Delta T}{\Delta e} = \frac{H}{LE}, \tag{6}$$

where Δe is the difference between measured vapour pressures at the considered levels. After that it was possible to calculate the latent (LE) and the sensible heat (H) fluxes from equations (1) and (6). The quality control of the calculations was based on Refs. 2 and 3. The difficulties in using the Bowen ratio method and its disadvantages are in detail described in Ref. 2.

3.3. *The Profile Method*

According to the similarity theory, the sensible (H) and the latent heat (LE) fluxes can be expressed through the dynamic (or friction) velocity (u_*), the dynamic temperature (T_*) and the dynamic specific humidity (q_*) as[4–6]:

$$H = -\rho_m c_p u_* T_* \quad \text{and} \quad LE = -L\rho_m q_* u_* . \tag{7}$$

Our calculations of the bulk Richardson number in finite difference form:

$$Ri = \beta \frac{\Delta T}{(\Delta u)^2} \Delta z, \tag{8}$$

where β is the stability parameter and Δu denotes the difference in wind speeds between the two considered levels, showed that it was possible to assume near neutral stratification (more than 80% of the absolute values for the bulk Richardson number were below 0.1). In the first approximation standard logarithmic profiles for wind speed (u), temperature (T) and specific humidity (q) could be used:

$$\frac{\partial u}{\partial z} = \frac{u_*}{\kappa z}, \quad \frac{\partial T}{\partial z} = \frac{T_*}{\kappa z} \quad \text{and} \quad \frac{\partial q}{\partial z} = \frac{q_*}{\kappa z}, \tag{9}$$

where K is the von Kármán constant ($K = 0.41$). From equation (9) the dynamic velocity (u_*), the dynamic temperature (T_*) and the dynamic specific humidity (q_*) were calculated as:

$$u_* = \Delta u \kappa \left[\ln\left(\frac{z_2}{z_1}\right)\right]^{-1},$$

$$T_* = \Delta T \kappa \left[\ln\left(\frac{z_2}{z_1}\right)\right]^{-1} \quad \text{and} \quad q_* = \Delta q \kappa \left[\ln\left(\frac{z_2}{z_1}\right)\right]^{-1}, \tag{10}$$

where $z_2 > z_1$.

3.4. *Causes of Uncertainties*

To prevent the acceptance of physically inconsistent or extremely inaccurate values it is necessary to reject values of the inappropriate data. The presented methods were applied after the physically established data filtering. The contradictory results obtained for the latent and sensible heat fluxes using the Bowen ratio and the profile method were supplemented by the data obtained from the energy budget measurements and calculations. In addition to the measurement errors, filtering and replacement are the main sources of uncertainties in calculations of evaporation. The influence of the filtering criteria selection for the calculated monthly average evaporation is 10–20%.

Further causes of uncertainties were our assumptions of equality of sensible and latent heat turbulent diffusion coefficients (used for the Bowen ratio method) and of the neutral stratification (used for the profile method). Finally, we noted that in the case of small gradients the Bowen ratio and the profile method were uncertain, too.[2]

4. Results

The diurnal variations of evaporation were calculated using the micrometeorological measurement dataset. Figure 4 shows the variations of the global (R_g) and the net radiation (R_n), the subsurface (S) and the latent heat fluxes obtained using the Bowen ratio method (LE_B) for a typical summer day with changing cloudiness (22nd August 2006). Figure 5 represents the comparison between the latent and sensible heat fluxes obtained using the Bowen ratio and the profile methods for the same day.

The monthly variations of evaporation were calculated from diurnal values. Figure 6 shows the monthly values of evaporation obtained by three different methods [the energy budget (LE_{eb}), the Bowen ratio (LE_B) and the profile method (LE_p), and demonstrates the uncertainties in the estimation of monthly evaporation values. Table 1 summarises the monthly values of evaporation, precipitation and inflow into the lake from September 2006 to April 2007. The studied period was very dry with extremely low precipitation in September 2006,

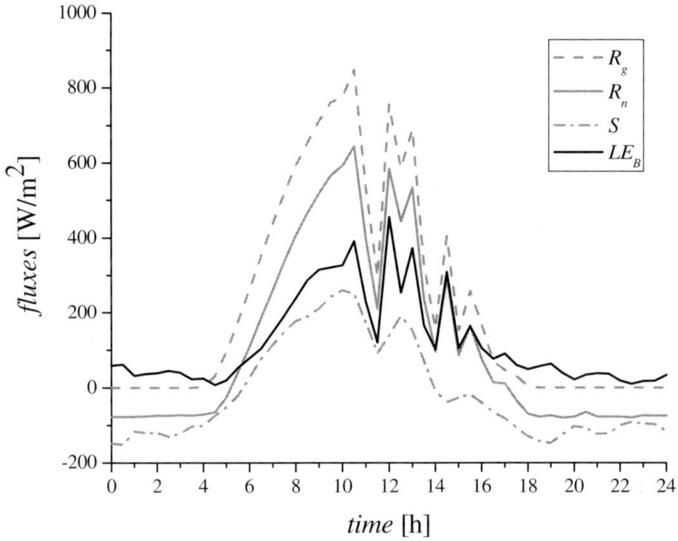

Figure 4. The diurnal variations of global (R_g) and net radiation (R_n), as well as subsurface (S) and latent heat fluxes (LE_B), calculated using the Bowen ratio method.

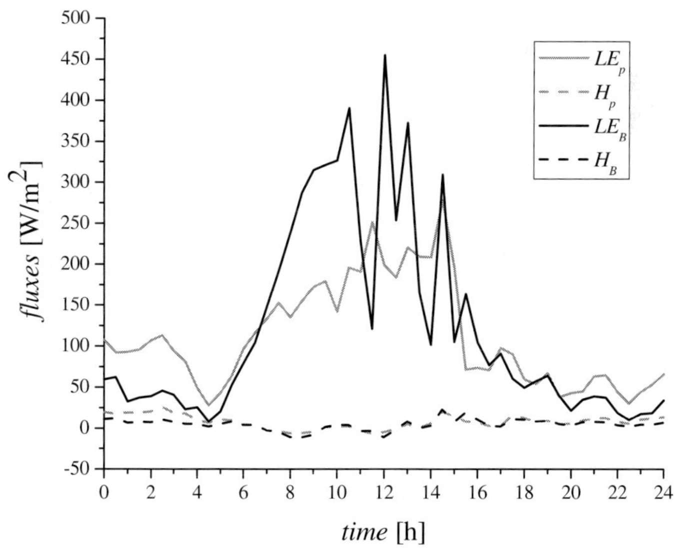

Figure 5. Comparison of the latent and the sensible heat fluxes obtained using the Bowen ratio and profile methods.

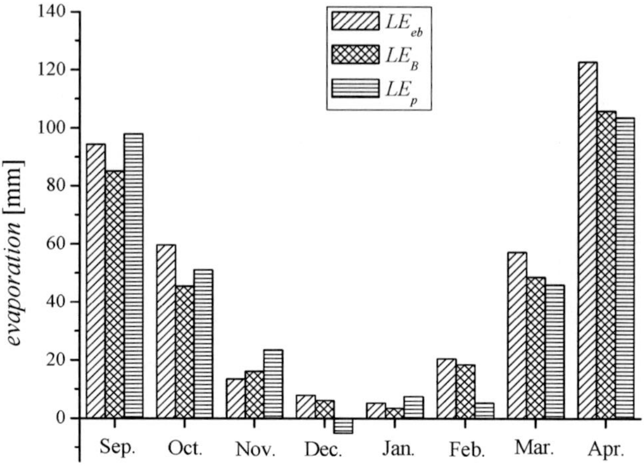

Figure 6. Uncertainties in calculations of monthly evaporation from September 2006 to April 2007.

Table 1. Uncertainties in calculations of monthly evaporation and inflow into Lake Kelemen-szék from September 2007 to April 2007

Month	$LE_{eb} = R_n - S$ [mm]	LE_b [mm]	LE_p [mm]	Precipitation [mm]	Inflow (min) [mm]	Inflow (max) [mm]
Sep.	94.4	85.0	97.8	1.0	2.0	15.4
Oct.	59.6	45.5	51.1	54.8	−30.3	−16.2
Nov.	13.5	16.1	23.4	31.0	−44.5	−34.6
Dec.	7.8	6.1	−5.1	7.8	−37.9	−25.0
Jan.	5.2	3.4	7.4	0.0	6.4	10.4
Feb.	20.4	18.4	5.3	27.9	21.4	38.7
Mar.	57.2	48.6	46.0	27.6	6.4	19.0
Apr.	122.7	105.8	103.6	0.0	23.6	42.7
Sum	380.8	328.9	329.6	150.1	−53.0	50.6

January and April 2007. During the studied period the water level decreased by 20 cm. Due to uncertainties in calculations of evaporation, the minimum and maximum possible inflow values were estimated.

Our calculations have proved our hypothesis about the weak connection between the Lake Kelemen-szék and groundwater flow system. The sign of the inflow (difference between precipitation, evaporation and water level) is changing during the year (Table 1). The water budget of the Lake Kelemen-szék is principally governed by evaporation and precipitation. This was confirmed after the end of the measurement period when the Lake Kelemen-szék ran dry in summer 2007.

5. Conclusion

The evaporation and the water budget of the shallow lake were determined by three different methods (the energy budget, the Bowen-ratio and the profile method), using micrometeorological measurements. For the studied period from October 2006 to April 2007 (except January 2007), the uncertainties in the calculations of monthly evaporation exceeded 10 mm. The obtained values and uncertainties in calculations of evaporation were to be expected.[7] The water budget of the Lake Kelemen-szék is above all governed by evaporation and precipitation, but our calculations have confirmed the hypothesis about the weak connection between the lake and the groundwater flow system.

Acknowledgements

The work was supported by the Hungarian Scientific Foundation (OTKA T-047159 Grant) and by the Hungarian National Research and Innovation Program (NKFP 088/2004 and NKFP 6/028/2005 grants). The second author expresses his sincere appreciation to the International Visegrad Fund for the financial support (scholarship) during the preparation of this work. The authors thank Ferenc Zsemle for his assistance during the measurements.

References

1. J. Tóth and I. Almási, *Geofluids.* **1**, 11 (2001).
2. A. Ohmura, *J. Appl. Meteorol.* **21**, 595 (1982).
3. H. Liu and Th. Foken, *Meteorol. Z.* **10**, 71 (2001).

4. T. Weidinger, J. Pinto and L. Horváth, *Meteorol. Z.* **9**, 139 (2000).
5. G.N. Panin, A.E. Nasonov, Th. Foken and H. Lohse, *Theor. Appl. Climatol.* **85**, 123 (2006).
6. D.T. Mihailović, S.T. Rao, K. Alapaty, J.Z. Ku, I. Arsenić and B. Lalić, *Environ. Model. Softw.* **20**, 705 (2005).
7. M.S. Mosner and B.T. Aulenbach, Comparison of methods used to estimate lake evaporation for a water budget of Lake Seminole, southwestern Georgia and northwestern Florida, *Proceedings of the 2003 Georgia Resources Conference* (2003).

CHAPTER 29

MODELLING AND IDENTIFICATION OF A PHARMACOKINETICS SYSTEM WITH TIME DELAY, WORKING EXAMPLE WITH DICLOFENAC DUO

MILAN R. RAPAIĆ AND MILENA PETRUJKIĆ

Faculty of Technical Sciences, University of Novi Sad, Novi Sad, Serbia

MILICA T. ATANACKOVIĆ AND JOVAN POPOVIĆ

Faculty of Medicine, University of Novi Sad, Novi Sad Serbia

In this paper we address the problem of modelling concentration-time profile of Diclofenac Duo drug in human blood after oral administration. We focus on linear-time invariant models with transport delay. The presence of delay forced us to use non-linear identification algorithms, genetic algorithm in particular. The problem was additionally challenging due to non-uniform sampling of data, as well as due to multiple peaks detected in profiles of certain patients. These problems were dealt with by means of approximation techniques and by superposition of lower order models. Responses of obtained models show very good resemblance to the measured data.

1. Introduction

In pharmacokinetics, mathematical modelling has historically played a vital role. Several model types have become quite common in the field, including linear compartment models, physiologically-based models and population models. However, several other promising modelling techniques have emerged recently. These new techniques rely on theory of dynamic systems, artificial neural networks, fuzzy logic and fractals. For a more detailed account of these techniques, we refer to[1] and references therein, and also to.[2–7]

The theory of dynamic systems, especially the theory of linear, time invariant systems, has been a valuable tool in engineering and other related fields during several previous decades. In the light of this theory, we consider a phenomenon of interest as a "black-box". Each such "box" remains in constant interaction with its environment. Influences of the surroundings to the system are identified as *inputs*, while the reverse ones are identified as outputs. Each activity in the *outputs* of the system is the consequence of activities in its inputs, as well as of the history of the system.

When the drug is administrated the result depends on many complex processes which occur in human organism. Therefore, the concentration-time profile is considered to be a crude surrogate of true consequence of absorption, distribution and elimination of the drug, body history and actual psychopathological state of the body. In contrast to the usual terminology in the dynamic systems theory, in pharmacokinetics the term "system" is used as a synonym for a body, or a body part. In the case considered within the present research, the drug input to the body is considered to be the input of the system and concentration of the applied drug in blood is the output. All the various processes inside the body are conveniently placed inside the "black box" and are consequently left aside, at least for the time being.

The use of dynamic system theory in pharmacokinetic modelling and analysis could provide several benefits. It could lead to better understanding of complex mechanisms in human organism, but also to therapeutic optimisation and individualisation of dosing of various drugs.

2. Description of the Experiment

Diclofenac Duo is a capsule with modified release which was orally administrated to healthy volunteers. Samples of blood were taken from these subjects before and during a 24-hour period after drug administration. In particular, blood samples were taken 15 min, 30 min, 45 min, 1 h, 1 h 20 min, 1 h 40 min, 2 h, 2 h 30 min, 3 h, 4 h, 6 h, 8 h, 10 h, 12 h and 24 h after the administration. Concentration of Diclofenac Duo was determined by means of HPLC (High Performance Liquid Chromatography).

3. Choosing the Appropriate Model Structure

The aim of the present paper is to model the concentration-time profile of the drug in human blood after its oral administration. In order to do so, we consider this profile as an impulse response of a specific single input-single output (SISO) linear time-invariant (LTI) system. If we could manage to identify the system properly, we would be able to reconstruct its responses. As it is well known, SISO LTI systems can be successfully described by differential equations, or equivalently by an operator called *transfer function*. In general form, the transfer function of a system reads:

$$G(s) = \frac{b_m s^m + \cdots + b_1 s + b_0}{s^n + a_{n-1} s^{n-1} + \cdots + a_1 s + a_0} \cdot e^{-s\tau}. \tag{1}$$

The Laplace operator is denoted by s. Numbers of poles and zeros are denoted by n and m, respectively. Transport delay is denoted by τ.

The process of identification consists of two steps. The first step is to choose the model structure and the second step is to estimate model parameters. The structure of a model is determined by degrees of numerator and denominator polynomials (m and n, respectively). There is no formal recipe for structure selection. In practice, the right structure is usually selected empirically, by trial and error. In general, degrees of the polynomials should be large enough to capture response dynamics, yet small enough to avoid over-fitting.

In the context of the present paper, the first-order and second-order models are considered. The transfer function of the first-order LTI system reads:

$$G(s) = \frac{b_0}{s + a_0} \cdot e^{-s\tau} = \frac{K}{sT + 1} \cdot e^{-s\tau}. \tag{2}$$

The transfer function of the second-order system reads:

$$G(s) = \frac{b_1 s + b_0}{s^2 + a_1 s + a_0} \cdot e^{-s\tau} = K \frac{sT_z + 1}{(sT_{p1} + 1)(sT_{p2} + 1)} \cdot e^{-s\tau} \tag{3}$$

The alternative parameters, K and T in (2) and K, T_z, T_{p1} and T_{p2} in (3) are introduced for convenience. Introduction of these parameters is common in engineering literature.

Since Diclofenac Duo has a modified release, many of the considered profiles are non-unimodal, they have more than one peak. Most of the time additional peaks are merely the result of measurement errors, but sometimes they reflect an additional, postponed drug action. In order to describe such non-unimodal behaviour, we assumed that overall profile is result of superposition of two profiles, each corresponding to a different drug component and each with its specific model. The overall transfer function reads:

$$G(s) = G_1(s) + G_2(s), \tag{4}$$

where G_1 and G_2 are transfer functions corresponding to the first and the second drug profile, respectively. Each of these functions is chosen either in the form (2) or in the form (3).

4. Estimation of Model Parameters

The second important step in modelling is parameter estimation. Most of the procedures for parameter estimation reported in literature, see for example, deal with discrete-time models, rather that with continuous-time models, such are (2) and (3). Therefore, in order to estimate parameters of continuous-time model, we must find its discrete-time equivalent first. There are several techniques reported in literature to be applied, impulse-invariant transform, for example.

Once the discrete-time system is obtained, its parameters are estimated by means of the Least Mean Squares (LMS) procedure. The LMS procedure is widely used, since it is linear and efficient. However, in the setting of the present paper it has several drawbacks. The first setback is that, in order to apply LMS in the first place, the delay must be known in advance. In practice, we try with different delays and adopt the most satisfactory one. In cases when there are two transfer functions with different delays connected in parallel, such as (4), this procedure is by no means convenient. The second problem is that LMS needs uniformly sampled data, which we do not have. The problem has been resolved by means of interpolation. Various interpolation algorithms were used, including piecewise linear interpolation and spline.

Another approach would be to use a non-linear algorithm that would directly minimise prediction error in time domain over the entire parameters set (including delay). Such an algorithm would not need *a priori* knowledge of delay and would not need uniformly sampled response profiles, since it would estimate parameters of continuous-time model directly. For example, the impulse response of a first-order system (2) is:

$$g(t) = \frac{K}{T} e^{-\frac{t-\tau}{T}} \tag{5}$$

The criteria to minimise would be:

$$J = \sum_{i=1}^{N} (g(t_i) - y_m(t_i))^2. \tag{6}$$

As it turns out, the minimisation of J is a non-trivial problem, since it is multimodal and highly sensitive to the initial guess (especially regarding the delay, τ). Therefore, it seemed natural to minimise it by means of global optimisation technique. Genetic algorithm was chosen since this is well known and highly successful global optimisation procedure. Particularly favourable feature of the global optimisation approach is that it can be used in exactly the same way, no matter how many responses we must deal in parallel.

5. Results

The concentration-time profile for Diclofenac Duo drug in human blood during 24 hours for a single patient is given in Figure 1(a). The modelled responses are shown in Figure 1(b). Actual measurements are

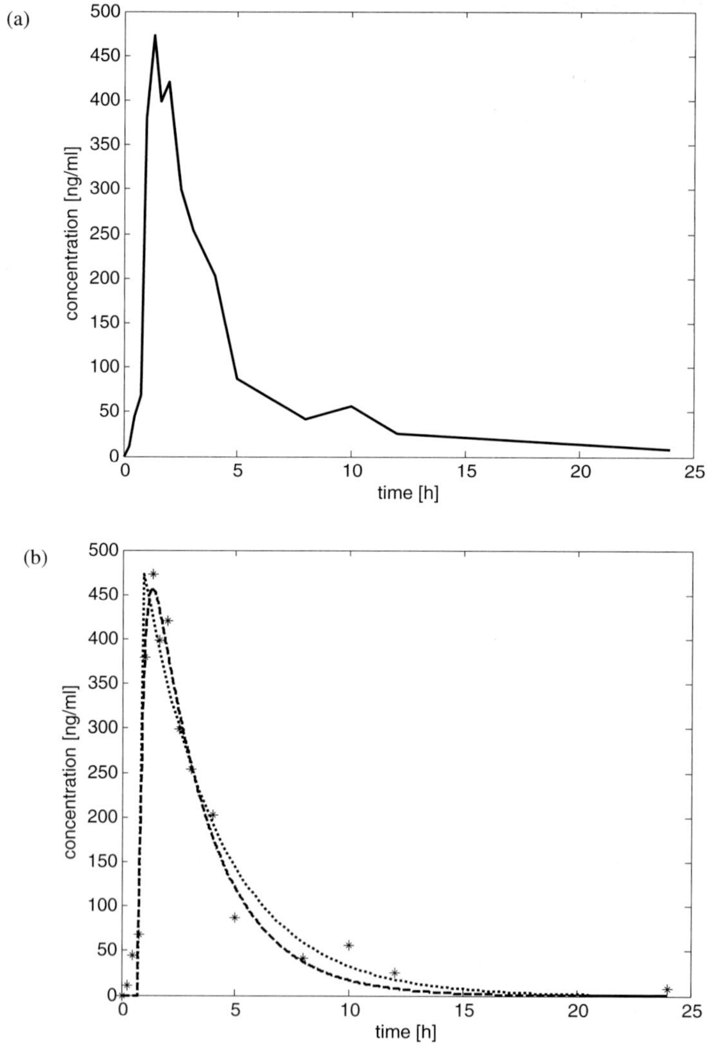

Figure 1. A sample concentration-time profile with one distinguishable peak.

Table 1. Parameters of obtained models

	First-order model	**Second-order model**
K	473.9366	1488.4
T_{p1}	3.3833	2.5809
T_{p2}	0	0.22603
T_z	0	0
τ	0.9585	0.72219

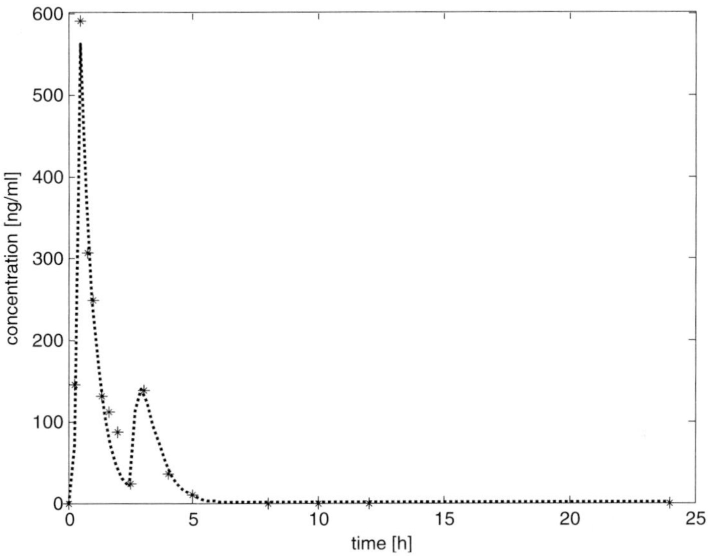

Figure 2. A sample concentration-time profile with two distinguishable peaks.

depicted by crosses (+), responses of the first-order model are drawn with dotted and responses of the second-order model with dashed lines. Model parameters are presented in Table 1. The profile presented in Figure 1 has no clearly distinguishable second peak. Figure 2 represents the profile with two separate maxima, together with the impulse response of the identified model. The original measurements are again denoted by asterisks, while the impulse response of the model is drawn with a dotted line. In this multimodal case we used

Table 2. Parameters of obtained model

	1st model	2nd model
K	460.74	159.52
T_{p1}	0.058767	5.4175
T_{p2}	0.58238	0.32451
T_z	0	0
τ	0.23687	2.4872

superposition of two second-order models. Specific parameters of these models are presented in Table 2.

6. Conclusion

The present paper demonstrates the applicability of identification and estimation techniques known from engineering in pharmacokinetic modelling. We have also demonstrated efficiency of these techniques, even when dealing with multimodal responses, and non-uniform data sampling. The genetic algorithm has been applied during parameter estimation. Our results demonstrate good resemblance between the responses of models obtained during identification, and actual concentration-time profiles obtained by HPLC measurement.

References

1. M. Durišova and L. Dedik, *Basic Clin. Pharmacol.* **96**, 335 (2005).
2. J. Popović, *Eur. J. Drug Metab. Pharmacokinet.* **23**, 469 (1998).
3. J. Popović, *Exp. Toxic. Pathol.* **51**, 409 (1999).
4. J. Popović, *Eur. J. Drug Metab. Pharmacokinet.* **29**, 205 (2004).
5. J. Popović, *Eur. J. Drug Metab. Pharmacokinet.* **31**, 79 (2006).
6. J. Popović, R. Mitić, A. Sabo, M. Mikov, V. Jakovljević and K. Đaković-Švajcer, *Eur. J. Drug Metab. Pharmacokinet.* **31**, 87 (2006).
7. J. Popović, *Eur. J. Drug Metab. Pharmacokinet.* **29**, 205 (2004).
8. K.J. Åström and B. Wittenmark, *Computer-Controlled Systems Theory and Design* (Prentice Hall, 1997).

CHAPTER 30

CHARACTERIZATION OF GAS/ PARTICLE PARTITIONING OF PCBS AND PAHS IN A PILOT AREA OF KRAGUJEVAC, SERBIA

MAJA TURK-SEKULIĆ, JELENA RADONIĆ AND MAJA ĐOGO

Faculty of Technical Sciences, University of Novi Sad Novi Sad, Serbia

Bombing and destruction of the industrial and military targets accompanied by incomplete combustion during the NATO operation in Serbia, caused emission of the persistent organic pollutants into the atmosphere, ground and underground water, sediment and soil. As the result of the air sampling campaign conducted in June/July 2004, the concentration levels of PCBs and PAHs in air samples in gaseous and particulate phase at the selected locations from the city of Kragujevac, Serbia, was determined using active sampling method and gas/particle partitioning of sixteen priority PAHs and seven PCB congeners was analyzed. GC/MS analysis was carried out in laboratories of Research Centre for Environmental Chemistry and Ecotoxicology (RECETOX), Masaryk University in Brno, Czech Republic. The relatively high concentration of PCBs was measured on locality inside the Zastava Car Factory (industrial locality), in proximity of the power generation building and two large transformers. From the residual levels of the PCB congeners in the air samples as gaseous and particulate phase it is evident that the most of the investigated congeners existed in the gaseous phase. Analysis of the gas and aerosol partitioning of PAHs in Kragujevac emphasized the presence of PAHs in both, gaseous and particulate phase. Gas/particle partitioning of PAHs showed uniform distribution at the investigated locations.

1. Introduction

Persistent Organic Pollutants (POPs) are a well-known class of chemicals subjected to long-range transport and bioaccumulation in humans and animals fat tissue and characterized by high resistance to environmental degradation processes.[6] POPs are toxic organic compounds of natural or anthropogenic origin and include substances such as polycyclic aromatic hydrocarbons (PAHs), dioxins, furans, polychlorinated biphenyls (PCBs) and organochlorine pesticides (HCH, DDT, HCB etc). Being capable of long-range transport, POPs released to the environment could travel through air and water (due to advection and diffusion processes) to regions far from their original source according to concentration, temperature and pressure gradient.

The removal of POPs from the atmosphere mostly depends on the gas/particle phase partitioning, since the deposition process of gases is different than for particles and it is obvious that the environmental fate of POPs in the environment is determined by the partition of the compound between different media such as air, water and soils, or between different phases of the same media. Distribution of the released pollutant between various environmental compartments depends on its numerous physical-chemical properties.

PAHs are primarily emitted into the atmosphere where they are mostly associated with particulate matter. However, PAHs are also found in the gaseous phase. PAHs having two or three rings are present predominantly in the vapour phase. PAHs that have four rings exist both in the vapour and in the particulate phase, and PAHs having five or more rings are found predominantly in the particle phase.

PCBs are dominant man-made organochlorines, which are banned due to their health risk to humans and to the environment. PCBs are extremely hydrophobic, highly resistant to chemical reactions and biological degradation, which leads to their persistence once introduced to the environment. PCBs were subsequently introduced into the environment and are now considered ubiquitous pollutants, found in most animal and human adipose samples, milk, sediment, air and numerous other matrices. The number and location of the chlorines (especially dioxin-like) govern both the environmental fate and toxicity of each congener.

During the NATO operation in Serbia, a huge amount of PAHs and PCBs was generated and/or released into the atmosphere, surface and groundwater, sediment and soil as the result of the bombing of industrial plants and installations, military targets and storage spaces filled with toxic chemicals and uncontrolled complete and incomplete combustion caused by explosions. After the air strikes, UNEP has assessed the consequences of the conflict for human settlements and the environment and defined four key hot-spots in Serbia: Bor, Novi Sad, Pancevo and Kragujevac. The air strikes in Kragujevac caused extensive environmental pollution, damaging soil, water and air. The main environmental problem was the leaking of 2500 kg of PCBs into the Morava River and contamination of groundwater and soil.[4]

In order to define the relevant phase into which chemicals primarily partition in air, the gas/particle partitioning of sixteen US EPA priority PAHs and seven PCB congeners was investigated and discussed in the paper. The obtained gas/particle distribution was compared with the values of fraction (%) of a compound associated with particles estimated using Junge-Pankow (JP) absorption model. The atmospheric levels of PAHs/PCBs used in the study were measured separately in the gas and particle phase of the atmosphere using active sampling method during the sampling campaign conducted in June 2004 at one of the hot-spots in Serbia, the city of Kragujevac, within the project APOPSBAL.

2. Materials and Methods

2.1. *Sampling locations*

Air samples were collected during the 24 hours for the period of three days (June 23–26, 2004) simultaneously using three HiVol samplers at the three selected highlighted localities in Kragujevac. Two samplers have been placed at the industrial area, inside the Zastava Car Factory: one at the power generation building (locality K1) and the other one at the main production building (locality K2). The third sampler was used as a background sampler located at the University of Kragujevac, city centre, (locality K3). Taking into consideration leaking of transformer oil during the air strikes and the permanent usage of the charcoal in the boilers, localities K1 and K2 are assumed to be the major source of the air pollution with PCB congeners and PAHs, while the motor vehicles at the locality K3 could be the source of PAHs emission.

2.2. *Sampling Methods*

Ambient air sampling was conducted in accordance with the U.S. EPA methods for the determination of toxic organic compounds in ambient air using the active sampling method with HiVol sampler containing a quartz fiber filter (which collects particle-bound compounds) and two polyurethane foam filters (which retain gaseous compounds).

2.3. *Sample Analysis*

PAHs and PCBs were analyzed by the laboratories of the RECETOX centre, Masaryk University, Brno, Czech Republic, using the GC/MS method for determining PAHs and GC/EC for analyzing PCB congeners.[1,5]

3. Results and Discussion

3.1. *Gas/Particle Partitioning*

Partitioning of semi volatile organic compounds between the gas and particulate phases in the atmosphere controls their rate of removal by wet and dry deposition, their atmospheric degradation and their long-range transport.[2,6,8,9]

Fractions of polycyclic aromatic hydrocarbons and polychlorinated biphenyls absorbed at the atmospheric particles, ϕ [%], at the localities K1, K2 and K3 are shown in Table 1.

Table 1. PAH/PCB fraction absorbed at the atmospheric particles, ϕ [%], at the localities K1, K2 and K3

Component	K1	K2	K3
Naphthalene	2.04	3.06	1.58
Acenaphthylene	2.00	0.76	0.64
Acenaphthene	3.37	4.29	7.89
Fluorene	0.24	0.29	0.52
Phenanthrene	0.22	0.22	0.31
Anthracene	0.64	0.37	0.49
Fluoranthene	2.61	2.36	1.23
Pyrene	4.69	3.12	1.46
Benz(a)anthracene	68.96	56.81	23.08
Chrysene	49.75	51.63	27.91
Benz(b)fluoranthene	95.79	97.55	94.84
Benz(k)fluoranthene	95.31	94.49	90.90
Benz(a)pyrene	98.52	100.00	100.00
Dibenz(a,h)anthracene	100.00	100.00	100.00
Indeno(123cd)pyrene	100.00	100.00	100.00
Benz(g,h,i)perylene	100.00	100.00	100.00
PCB 28	0.11	0.39	2.86
PCB 52	0.21	1.70	7.69
PCB 101	0.85	2.74	8.69
PCB 118	1.43	7.10	11.11
PCB 138	2.62	9.09	9.09
PCB 153	1.64	5.74	10.00
PCB 180	6.30	17.50	0.00

The most often used model for describing the gas/particle partitioning of the persistent organic pollutants was proposed by Junge-Pankow (JP).[3,7] According to the JP model, the fraction of a compound associated with particles, ϕ, is calculated using vapor pressure of the sub cooled liquid at the ambient temperature, p_{0L}, available aerosol surface, θ, (m^2 particles per m^3 air) and parameter c (Pa m) that depends on the difference between the heat of desorption from the particle surface and the heat of vaporization of the compound and the surface properties of the suspended particles, $\phi = c\theta/(p_{0L} + c\theta)$.[6]

For the estimation of the fraction ϕ of the investigated compounds, the sub cooled liquid vapor pressure at the average temperature during the three days of 24 hours sampling periods was used. Calculations were made assuming c = 0.172 Pa m^3, and $\theta = 1.1 \times 10^{-3}$ m^2 m^{-3}, which are representative of aerosols from urban areas. Comparison of the observed and estimated PAHs/PCBs fractions absorbed by atmospheric particles, see Figures 1 and 2.

Relative deviations of estimated ϕ values using JP approach to the calculated values obtained from the measured concentration levels of polycyclic aromatic hydrocarbons and polychlorinated biphenyls at the localities K1, K2 and K3 are shown in Table 2.

3.2. *Discussion*

PAHs investigated in this study range from those which generally can only be detected in the vapour phase (such as 2 or 3-ring PAHs) to compounds which can be detected only in collected particles (5 and 6-ring PAHs). Gas/particle partitioning analysis shows that while Flo, Phe and Ant were collected almost completely on the polyurethane-foam filters (vapour absorbent medium), B(b)F, B(k)F, B(a)P, D(a,h)A, I(123cd)P and B(g,h,i)P could only be detected in the glass-fibre filters at all localities (Table 1). Results of the gas/particle distribution behaviour analysis emphasize presence of PCB congeners mainly in the vapour phase (Table 1).

Analyzing the relative deviation of estimated ϕ values using JP approach to the calculated values obtained from the measured concentration values, see Figures 1, 2 and Table 2. Excellent agreement is noticed at the high molar weight PAHs in the atmosphere, as well at the Pyr and B(a)A at the locality K3. Relative error at B(b)F, B(k)F, B(a)P, D(ah)A and B(ghi)P ranges from 0.048746 to 10.69297%; at Pyr and B(a)A relative error amounts to 4.92% and 2.61%, respectively. The highest relative deviations of estimated/calculated ϕ values are detected at all investigated PCB congeners (relative deviations of ϕ ranges from 6.68% to extremely high values) and at the low molecular weight PAHs — Nap, Acy and Ace (up to 99.98%) and in the case of Chr, at the city centre (127.82%). On the

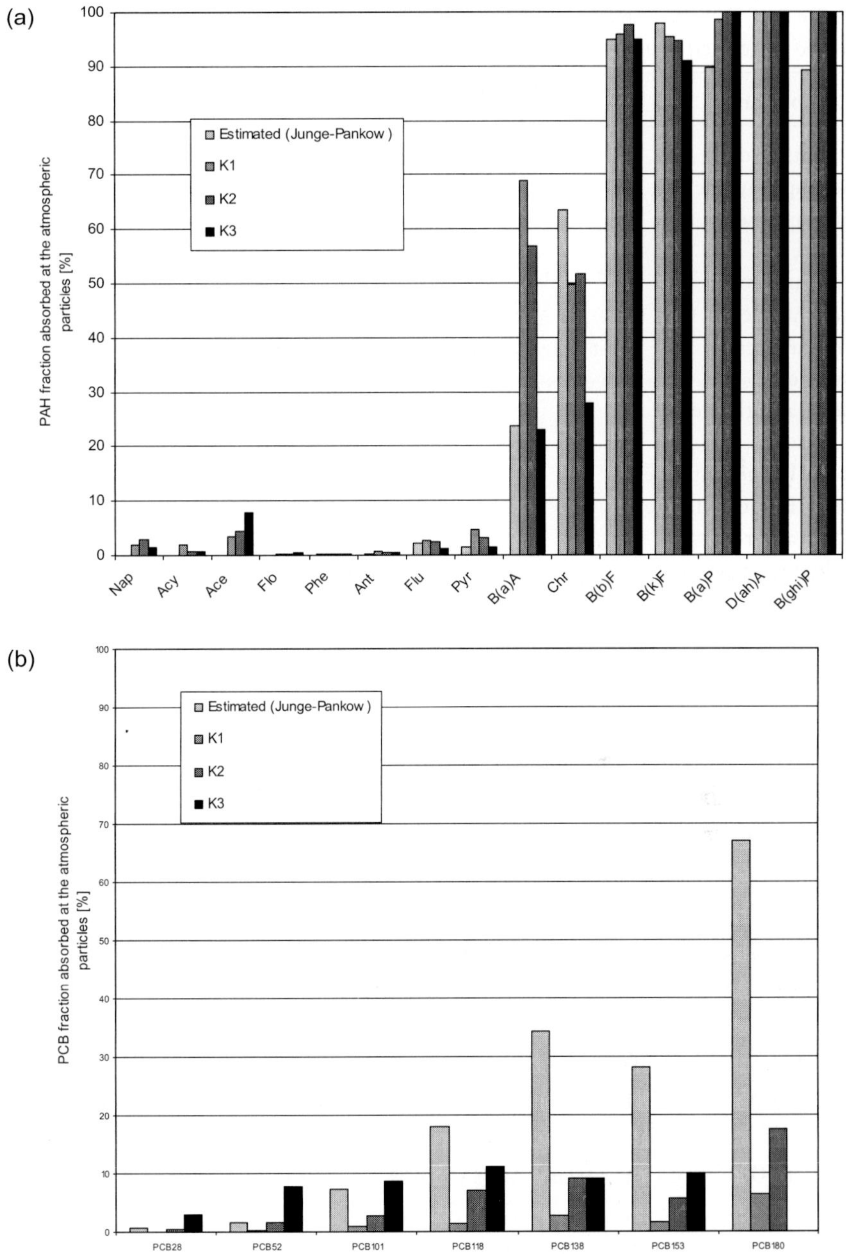

Figure 1. Comparison of the estimated and calculated (a) PAH and (b) PCB fractions absorbed by atmospheric particles.

Table 2. Relative deviations of estimated ϕ values using Junge — Pankow approach to the calculated values at the localities K1, K2 and K3 [%]

Component	K1	K2	K3
Naphthalene	99.97	99.98	99.97
Acenaphthylene	99.78	99.41	99.30
Acenaphthene	99.63	99.71	99.84
Fluorene	89.00	90.89	94.92
Phenanthrene	24.37	24.37	46.33
Anthracene	62.45	35.04	50.95
Fluoranthene	19.47	10.94	70.88
Pyrene	67.34	50.90	4.92
Benz(a)anthracene	65.66	58.31	2.61
Chrysene	27.81	23.15	127.82
Benz(b)fluoranthene	1.02	2.81	0.03
Benz(k)fluoranthene	2.68	3.57	7.66
Benz(a)pyrene	8.91	10.26	10.26
Dibenz(a,h)anthracene	0.05	0.05	0.05
Benz(g,h,i)perylene	10.69	10.69	10.69
PCB 28	581.25	92.15	73.80
PCB 52	655.47	6.68	79.37
PCB 101	759.37	166.59	15.94
PCB 118	1152.58	152.28	61.22
PCB 138	1212.02	278.16	278.16
PCB 153	1614.65	389.90	181.20
PCB 180	963.89	283.00	not defined

other hand, findings obtained by the other authors cite that the vapour/particle partitioning of the lighter (low molecular weight) PAHs can be predicted using JP model within the reasonable limits.

Taking into consideration toxicological relevance of B(a)P, PAHs are often represented by its concentration levels. Detected concentrations of B(a)P (PUF + GF) ranged from levels near the detection limit (about 0.01 ng/m^3) from air samples detected at locality K3 to levels of 0.135 (locality K1) and 0.158 (locality K2). Levels of B(a)P at the sites K1 and K2 are above immission limit value for the B(a)P regulated in Serbia (0.1 ng/m^3). Concentration levels of other PAHs recognized by as carcinogen (B(a)A and D(a,h)A) are near the detection limit for all investigated locations. At the time of the air sampling campaign, concentration levels of B(a)P, as the representative of PAHs, were measured only in Belgrade, by the Institute of public health. Obtained values demonstrate presence of B(a)P above immission limit value at all the field sites and permanent exposure to this carcinogenic pollutant.

At the investigated locations, the highest total PAH concentration value was measured at the locality K2, ΣPAH = 34.309 ng/m^3. Total PAH concentrations at the other two locations amounted to ΣPAH = 26.409 ng/m^3 and ΣPAH = 19.382 ng/m^3, respectively. The most abundant component at all localities was Phe (20.933 ng/m^3 at the locality K2), while D(ah)A was present at the least extent. Incomplete combustion in the coal furnace seems to be the major source of Phenantrene in the Zastava Car Factory.

PAHs with two and three rings constitute the major fraction of atmospheric PAHs for all three field sites. The fraction of PAHs with four rings in the atmosphere is neglected for the locations K1 and K2, while it is near the fraction of PAHs with two and three rings at the site K3, the locality with the lowest overall PAH concentration. For all field sites, the contribution of PAHs with five and six rings in the atmosphere is insignificant. Similar concentration trends of the investigated PAHs point out that detected levels result from the permanent incomplete combustion of organic matters at the selected sites, global atmospheric transport and persistence, without presence of the significant and specific source of pollution.

As the direct result of NATO operation in Serbia, high concentration levels of PCB congeners in ambient air are detected at the selected localities of the investigated region. Summary concentration of the seven PCBs at the locality K1 has reached the value of 40.149 ng/m^3. Locality K1, placed in the vicinity of the big transformer using pyralene, with the highest detected PCBs level presents the important source of pollution. High concentration values of the PCBs could be an adequate sample for the analysis of the characteristic contribution of the 7 EPA congeners. Exceptions in concentration levels between industrial and urban areas pointed out proper position of the factory block with the direction of the airstreams opposite to the city residential district.

PCB 52 and 101 constitute the major fractions of the total PCBs at the locality Kl indicating the major source of contamination in the area, while the fractions of PCB 28 and 180 are neglected. At the field sites K2 and K3, the similar concentration trend of individual PCB congeners with the major fraction of PCB 28 is noticed.

4. Conclusion

GC/MS analysis of the air samples shows the presence of residual quantities of PCBs and PAHs in air samples from three localities in Kragujevac. Since vapour pressure value is considered to be the dominant physical-chemical property of the compounds which influence the vapour/particle partitioning, obtained distribution of the investigated pollutants is as expected. Initial findings suggest that the gas/particle partitioning of the high molar weight examined PAHs is in agreement to the JP absorption model. Extremely high levels of PCB congeners result from the damaging of transformers during the bombing of the industrial complex in Kragujevac. Similar concentration trends of the tested PAHs point out that detected levels result from the permanent incomplete combustion of organic matters at the selected sites.

5. Recommendation and Perspectives

Determination of the gas/particle partitioning behaviour of the selected POPs is of extraordinary importance for general estimates of wet and dry deposition from the atmosphere, modelling of the environmental fate and transport phenomena through the atmosphere and other environmental compartments and exposure of human population to these very toxic pollutants at micro, mezzo and macro level.

References

1. Assessment of the selected POPs in the atmosphere and water ecosystems from the waste generated by warfare in the area of formal Yugoslavia APOPSBAL: EU Project ICA2-CT2002–10007 (2005).

2. T. F. Bidleman, *Environ. Sci. Technol.* **22**, 361 (1988).
3. C. E. Junge, *Basic Considerations about Trace Constituents in the Atmosphere as Related to the Fate of Global Pollutants.* Suffet IH Ed. (Wiley, 1977).
4. S. Kaisarević, N. Andrić, S. Bobić, J. Tricković, I. Teodorović, M. Vojinović Miloradov and R.Z. Kovačevic, *Bull. Environ. Contam. Toxicol.* (2007).

CHAPTER 31

IMPORTANCE AND POSSIBILITIES OF CORN PRODUCTION IN HUNGARY IN THE FUTURE

DANIELLA BOKSAI AND ÉVA ERDÉLYI

Faculty of Mathematics and Informatics, Corvinus University of Budapest, Budapest, Hungary

Climate change has a major influence on our living in many ways. Agriculture can help in reducing carbon dioxide quantities in the atmosphere by increasing its absorption and decreasing its emission. Energy potential of Hungary is approximately half-half of products and by-products. Corn is an energy crop with significant by-production. Using by-products as green dung and for biogas production can be a step in prevention of global warming. We analysed the effects of changing climate on the proportion of grain, leaf, corn-stalk and root in corn biomass by using the 4M model for different climate scenarios as weather inputs.

1. Introduction

Global warming is one of the biggest problems in the world today. The main reason for global temperature increase is the increase of carbon dioxide concentration in the atmosphere.[1,6,7] The major cause of this increase is continued combustion of fossil fuels. Agriculture can help reduce global warming by increasing the absorption (land use,[4] irrigation, plant improvement) and decreasing the emission of carbon dioxide (biomass as energy resource[3,8]). There is a wide scientific consensus that if these changes continue, significant damage to global ecosystems, food production and economies will ensue. We analysed the possibilities that corn production can provide for bio-energy production.[9,10] Energy potential of Hungary is approximately half-half of products and by-products.[5] Corn is one of the most cultivated plants in Hungary. It is an energy crop with significant by-production. Using by-products for bio-fuel production and as green dung can be a step in prevention of global warming.

2. Bio-Fuel and Biomass

"Bio-fuels" is a term that commonly denotes liquid or gaseous fuels made from biomass. The biomass can come from different sources: starch from cereals, grains and sugar crops, waste products from agriculture and forestry, etc. The advanced bio-fuels from biomass can be produced using several different technical processes; a wider variety of potential feedstock would be available such as crops residues. Agricultural residues include a wide range of plant material produced along with the main product of the crop. Examples of biomass for energy production could be cereal straw, orchard pruning or corn stems. Producing biogas is also a sustainable technology. Biogas-produced from corn, for instance — as a renewable fuel can be used to power electricity generators, to provide heat and produce a soil improving material. Europe is facing increasing efforts for bio-energy production support, both based on the existing resources and energy crops implementation. There are biomass-power projects in Hungary as well — in Pécs, Borsod, Tiszapalkonya and Ajka. Debrecen is a city where local buses are already operated by bio-fuel. Heat and electricity from bio-based origin, together with transportation fuels production will soon start strong competing demands for solid biomass origin supply. Therefore, the assessment of biomass resources supply and future biomass potentials is of great importance.

3. Materials and Methods

3.1. *Scenarios*

Climate scenarios can be defined as relevant and adequate pictures of how the climate may look in the future. During our research, we applied the most commonly accepted scenarios presented in international reports, such as:

- BASE scenario, which is the base for all other scenarios with the parameters of present days
- Two steady double carbon dioxide scenarios created by the Geophysical Fluid Dynamics Laboratory (USA), GFDL2535 and GFDL5564, concerning the years around 2030 and 2060 respectively
- UKHI and UKLO (high and low-resolution equilibrium) and UKTR (high-resolution transient climate change experiment) worked out by the United Kingdom Meteorological Office (UKMO).

For scenario generation, the so-called GCM-s (General Circulation Model or Global Climate Model) was used. In this work, the GCMs were downscaled to Debrecen, the basic object of our calculations, as an important centre of agricultural production in Hungary. The scaling of the scenarios for the region in Hungary was made in the frame of CLIVARA project (Climate Change, Climatic Variability and Agriculture in Europe).[6] These meteorogical data include daily average, minimum and maximum temperature, precipitation amount, radiation values.

3.2. *4M Model*

In our modelling research we used the 4M model, which has been developed by the Hungarian Agricultural Model Designer Group from the various institutes in the country. It contains several models to describe the physiological interactions of soil — plant systems and offers a possibility of building up different system models in it for the specific purposes of the users need. The CERES model was chosen to be a starting point and was adapted to Hungarian circumstances. The simulations were run for the daily average temperature, precipitation amount and radiation forecasted by climate scenarios.

4. Modelling Corn Biomass for the Future

Living under changing climate conditions we need to prepare for the future. Energy potential of Hungary is approximately half-half of products and by-products. Corn is an energy crop with significant by-production. We analysed the impacts of climate change on the growing periods of corn by using the 4M model[3] for different climate scenarios as weather. We analysed the effects of changing temperature on the proportion of grain, leaf, corn-stalk and root in corn biomass by using the 4M model for different climate scenarios as weather inputs. Debrecen was the centre of our research, as one of the most important centres of agricultural production in our country. We analysed the impacts of climate change on the growing periods of corn by using the 4M model for different climate scenarios. We examined the effects of changing meteorological conditions on the proportion of corn-stalk and leaf mass in biomass, as well. Using biomass as a substitute for fossil fuel is highly prioritised, but the primary aim of agricultural production is food and feed supply. Converting secondary biomass, plant residues to valuable energy products might be a solution. That is why the secondary biomass quantities are discussed, too.

4.1. *Modelling Corn Secondary Biomass*

The results obtained for the UKTR scenario showed no significant differences in relation to the BASE scenario, but, the results for GFDL2 predicted smaller and for UKHI the smallest values. The results of the GFDL5 scenario showed significant increase in the predicted quantities of corn secondary biomass as compared to the BASE scenario. On the average (Table 1), UKHI showed the smallest values, GFDL2 and UKTR predicted a little lower quantity while GFDL5

Table 1. Predicted average quantities for bio-fuel production

Scenarios	BASE	GFDL2	GFDL5	UKHI	UKLO	UKTR
Biomass (kg/ha)	4977	4254	5538	3842	5727	4665

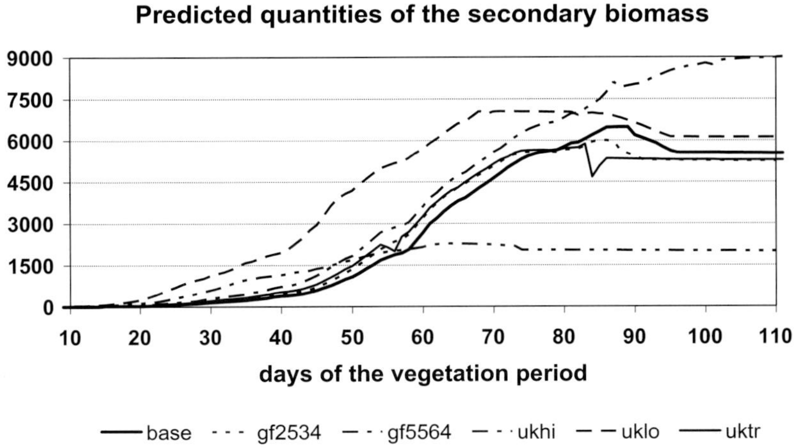

Figure 1. Predicted quantities of the corn secondary biomass, comparing six different climate scenarios for a randomly given year.

and UKLO showed higher values compared to the results of the BASE scenario (Table 1).

Since scenarios are given for 31 independent years (except for the UKTR scenario), it was possible to analyse them statistically. The obtained results indicated that the UKTR scenario showed no significant difference in relation to the BASE scenario, while the result for UKLO was very similar to the result of the GFDL5 scenario, showing significant increase in the predicted quantities of secondary corn biomass compared to the scenario BASE, but GFDL2 and UKHI showed lower values, see Figure 1.

4.2. *Modelling Corn Phenology and Biomass*

Considering considerable risk increase in corn production,[2] we decided to analyse the future corn production using a crop model and the same climate scenarios. We wanted to learn corn yields to be expected in the Debrecen region. For analysing the biomass changes we used again the 4M model for different climate scenarios mentioned above, as weather inputs. According to the results, we can say that moderate warming could increase the yield, but, on the other hand, further temperature increase can cause big losses. We also showed that the variability did not change much for the used climate scenarios, whereas the estimated mean yield did.[5]

Analysing the climate scenarios for the same region in the future it can be seen that the sum of the temperature averages increased over time in the past and the scenarios predicted even more drastic increases. This means that the starting date (sowing), ending date (ripening) and the length of the phenophases of growing plants will change, too. Examining the changes in the phenological phases of the plant we can see that climate change has the effect of shortening the length of the growing periods and, moreover, their starting dates are shifting to earlier dates.

We compared the sums of precipitations for the growing season of crops, as well. They showed a slow decrease of the accumulated precipitation for the examined periods in the region of Debrecen as well as smaller values in the future predicted by the scenarios. They also showed a great variability in the amount of precipitation, which means that the frequency of extreme weather events, such as droughts and floods are more probable.

The next step was to test the efficacy of three drop irrigation systems. The location of our experiment was the same, Debrecen region. We used the same 4M model for climate scenarios mentioned above as weather inputs. Our standard was BASE, which represents the current weather conditions. Biomass and grain levels decreased in the case of GFDL2534 and UKHI, compared to BASE. When comparing runs without treatment and runs with irrigation, we found that biomass and grain levels increased significantly in all scenarios. Our results have shown that climate change itself does not increase biomass or grain levels, while significant increase in yield can be reached by irrigation. This probably means that economic aspects regarding not using watering are worth reconsidering.

4.3. *Modelling Corn Biomass Proportion*

The results obtained for the proportion of the parts of the plant of the GFDL2 scenario were very similar to the results for the GFDL5 and the results for the UKTR showed no significant difference as compared to the BASE scenario. Next, we presented the two most different results, the Figures for the BASE and GFDL5 scenarios, see Figures 2 and 3.

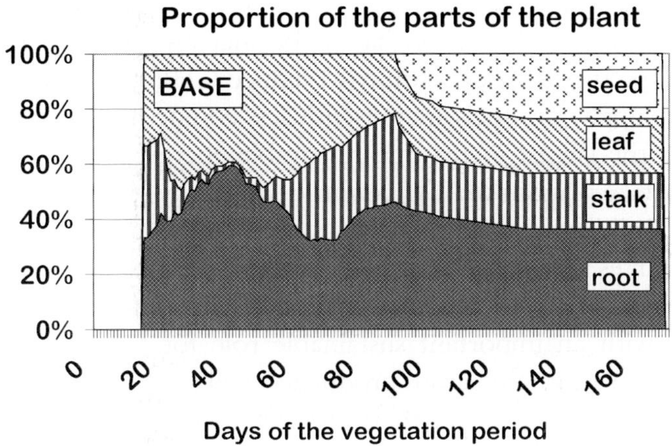

Figure 2. Biomass proportions of the parts of the corn plant simulated by climate scenario BASE.

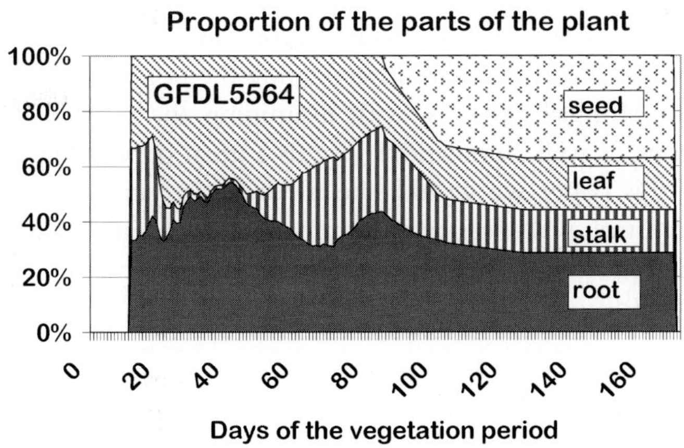

Figure 3. Biomass proportions of the parts of the corn plant simulated by climate scenario GFDL5564.

5. Discussion and Conclusion

Recently released IPCC WGI fourth assessment report illustrates human influence in warming impact on the global climate. Nowadays, the world energy supply is dominated by fossil fuels. Biomass resource sectors, such as agriculture, are playing a very significant contributing role. In order to accomplish the Kyoto greenhouse gas (GHG) reduction targets, the current utilisation of biomass has to be increased rapidly. Bio-energy gives Europe the best opportunity to reduce GHG emission and secure its energy supply. Bio-fuels are presently less competitive than fossil fuels. However, the largest increase in renewable energy use, in the coming years, will take place in the EU countries driven by strong governmental support. Building the new energy structure based on different bio-energy sources should be the main target, where beside wind, solar and hydro energy, biomass becomes an integral part of the overall energy strategies with an important sustainable role for bio-energy and bio-refineries to play. Energy crops production will have to compete for land areas with other crops, and their production has been regarded as an interesting issue in most European countries. Research and field trials with different energy crops species have been conducted to estimate yield, growth rate, survival rate, harvesting technology, etc. Secondly, their role in climate protection is in systematically sequestering carbon dioxide in soils and in plant biomass. Most promising crops for certain agro-climatic conditions have been selected already in some countries. We continue our research with investigating the potentials of agriculture residues in Hungary, that is, with examining the potentials of other energy crops, as well. Simulations give us a great opportunity.

With this paper we would like to draw attention to the importance of creating well-designed descriptive-forecasting systems, as well as defining the optimum preparatory and response strategies to the changed conditions.

Acknowledgements

Our work was supported by OTKA T042583 tender and the Jedlik Ányos NKFP6-00079/2005 program.

References

1. E.M. Barrow and M. Hulme, Constructions of scenarios of climate change and climatic variability: Development of climate change scenarios at a range of scales, Climate change, Climatic Variability and Agriculture in Europa, Annual Report, Environmental Changa Institute, University of Oxford, 13 (1996).
2. É. Erdélyi, A. Ferenczy and D. Boksai, Climate change and cereal crops crowing in Hungary, *EFITA Conference*, Glasgow, Scotland (2007).
3. N. Fodor, G. Máthéné-Gáspár, K. Pokovai and G.J. Kovács, *European J. of Agr.* **18**, 389 (2002).
4. M. Hoogwijk, A. Faaij, B. Eickhout, B. de Vries and W. Turkenburg, *Biomass Bioenerg.* **29**, 225 (2005).
5. L. Horváth and É. Erdélyi, How climate change influences the field crop production, *ECO-Conference*, Novi Sad, Serbia (2006).
6. The Physical Science Basis; Summary for Policymakers, IPCC Climate Change, The IPCC 4th Assessment Report (2007).
7. J.T. Houghton, L.G. Meira Filho, B. Callander, N. Harris, A. Kattenberg and K. Maskell, *IPCC Climate Change 1995, The Science of Climate Change* (Cambridge University Press, 1996).
8. S. Kim, and B.E. Dale, *Biomass Bioenerg.* **26**, 361 (2004).
9. M. Ladányi and É. Erdélyi, The increase of risk in corn production detected by a new stochastic efficiency method, *Agrárinformatika*, Debrecen (2005).
10. M. Ladányi, Alternatives of process approaches in agro-ecosystem modeling, Ph.D. Thesis, Corvinus University of Budapest (2006).

PART 4

HUMANITY ISSUES

CHAPTER 32

ENVIRONMENTAL POLICY MEASURES AND STAKEHOLDERS AT THE LOCAL COMMUNITY LEVEL — CASE STUDY OF NOVI SAD

JOVANA ČIKIĆ AND ŽIVOJIN PETROVIĆ

Faculty of Agriculture, University of Novi Sad, Novi Sad, Serbia

Environmental protection has a great impact on the overall development (whether regarding global society or local community). Therefore, it implies an organised social activity. Environmental preservation and improvement mean the defined goals, priorities and measures of environmental policy, as well as stakeholders' network participating in their fulfilment. This paper presents the results of research within the project titled "Participatory Environmental Assessment for Developing Local Environmental Action Plan (LEAP) of the City of Novi Sad". Results show statements of interviewed citizens on measures of local environmental policy and stakeholders who participate in environmental policy actions.

1. Introduction

Environmental protection is one of three pillars of development (next to the economic and social components). Along with the change in theoretical, but also practical approach in addressing environmental issues,[a] a greater deal of attention is paid to social and political context of environmental reforms, especially to the enforcement of environmental policy measures and performances of the stakeholders' network.

[a] The change is about the turn from a technically oriented approach ("end-of-pipe") to a more socially and culturally sensible one.

In the context of urban community growth and development, environmental protection (in the form of local environmental policy) has even greater significance. Urban areas are loci of people concentration, therefore, concentration of diverse social functions and activities with actual and potential impact on the environment. Environmental policy, at the community level, represents an intersection of global environmental policy goals, priorities and measures and, at the other end, local social circumstances and specific environmental problems.

According to the bottom-up approach in decision-making and policy planning processes and public right to know,[b] participation of stakeholders, especially local inhabitants, is considered vital to environmental policy establishing and, its implementation. One of possible ways to include local community population in policy planning is a participatory environmental assessment, as a constitutive element of LEAP[c] drafting (Local Environmental Action Plan). Apart from enabling data collection, LEAP is a good way to increase public interest in environmental issues. Public participation in environmental assessment is a foundation for more active role in policy measures enforcement.

2. Methodology

Participatory environmental assessment[d] was based on a sample of total population in four spatial units: Novi Sad, Petrovaradin, Sremska Kamenica and Veternik. The survey included 98.1% of a predefined

---

[b] See the basic principles of "Convention on Access to Information[1]: right to be informed on environmental issues, right to participate in the decision-making process, right to protect the previously listed rights.

[c] LEAP is, at the same time, the environmental decision-making process and the final document of such process — a local environmental strategy.[3] Environmental assessment is one step in LEAP drafting, which can be expert or joint (participatory) environmental assessment.[2]

[d] Participatory environmental assessment was conducted within the project titled "Participatory Environmental Assessment for Developing Local Environmental Action Plan (LEAP) of the City of Novi Sad" (2005/2006.). The research team from the Faculty of Agriculture (Novi Sad) participates in project implementation. The Ordering party was the Department for Environmental Protection of the City of Novi Sad.

population sample (1000). Such a large scope of a realised sample indicates that the population is very interested in environmental issues. The survey was based on a questionnaire containing 51 questions divided into five groups.[e]

3. Results

Results of participatory environmental assessment show that local inhabitants of Novi Sad are deeply concerned about environmental issues, especially because of the consequences of environmental degradation and deterioration. Therefore, they are interested in options suitable for resolving of environmental problems. Participatory assessment gave the opportunity to the citizens to evaluate priorities of environmental measures. Locals formed their opinion on each environmental measure's priority based on two criteria: one is the evaluation of efficiency of each environmental measure and second is his/her evaluation of a social context as the framework for defining and fulfilling of such measure. The indicator of the overall priority is a number of citizens that consider each measure to be at the high priority level. The participants were offered nine previously defined measures[f] divided into four categories:

1. Technological measures ("end-of-pipe" measures) — install purification filters; complete replacement of old, dirty technologies with new, modern and improved ones,
2. Economic measures — implementation of "polluter pays" principle; increase of "user pays" taxes,
3. Organisational and institutional measures — development of local environmental strategy, efficient environmental supervision, enabling citizens to participate in decision-making process concerning environmental issues,

[e] Five data groups: demographic data on a respondent, environmental assessment of the nearby space for living (residential block), environmental assessment at the local community level, public participation in environmental decision-making and practicing, environmental information etc.

[f] Each citizen was also given an option to define the measure for which he/she believes that was left out.

4. Environmental information and education — implementation of environmental education, increscent of the scope and quality of the environmental information (in media).

According to the participatory assessment, the first group of measures has the highest priority among the listed four groups. Such result is expected, due to the still dominant technological orientation in our environmental protection policy. In addition, the "end-of-pipe" approach is the most noticeable one and public is likely to be the most fascinated with new technical and technological solutions. Although the "end-of-pipe" approach is necessary, it is simply not sufficient in resolving environmental problems.

Within the first environmental policy measures group, the first measure, i.e. installation of purification filters, has a higher priority than the second one (complete replacement of old, "dirty" technologies with new, modern and improved ones). Such result is based on each respondent's assessment of possibility for each measure to be fulfilled in the present social context. Citizens who participated in environmental assessment considered that complete replacement of "dirty" technologies is not feasible, considering the financial costs and polluters' readiness to completely re-design the production process.

Within the group of economic measures, the citizens who participated in environmental assessment gave the higher priority to implementation of the "polluter pays" principle. The existing environmental laws already define such measure, but up to now, it has not been adequately enforced. Although the "polluter pays" principle seems to be a restrictive measure, it is, from the long-term perspective, a positive measure aiming to reform environmental behaviour of those whose activities have destructive effects on the quality of the environment. However, participatory environmental assessment showed completely different results for the second economic measure (increase of "user pays" taxes). The assumption is that there are four main reasons that determined such public opinion:

- Non-acceptance of the shared responsibility concept — participants in environmental assessment think that everyone has to take responsibility for their actions;

Table 1. Citizens' opinion on environmental measures priority

Measure	Low	Medium	High	Not respond	Total
	Level of priority				
A. Technological measures					
1. Install purification filters	1.1	8.7	87.6	2.7	100.0
2. Replace "dirty" technologies with new, modern and improved ones	1.1	15.8	79.4	3.7	100.0
B. Economic measures					
3. Implementation of the "polluter pays" principle	2.8	13.0	82.0	2.2	100.0
4. Increase of "user pays" taxes	19.3	38.8	31.5	10.4	100.0
C. Organisational and institutional measures					
5. Efficient environmental monitoring	3.7	14.7	71.9	7.1	100.0
6. Enabling citizens to participate in environmental decision making	8.4	26.6	62.9	2.1	100.0
7. Establishing of local environmental strategy	2.7	28.4	61.4	7.4	100.0
D. Environmental information and education					
8. Environmental education	7.3	31.8	57.8	3.2	100.0
9. Increase of scope and quality of environmental information	6.5	34.5	54.9	4.1	100.0

- The measure is considered as another attack on a family budget;
- It is more likely that participants (public, in general) are not familiar with the rate of "user pays" taxes;[g]

[g] The "user pays" tax has been collecting since 1995, pursuant to the City Council Decree. It rates only 0.10 RSD/m^2 per households or 0.18 RSD/m^2 per business buildings.

- Public shows no confidence in local government institutions, they are not sure that their money from the "user pays" taxes is being spent in the appropriate way.

In the group of organisational and institutional policy measures, the participatory environmental assessment showed that higher priority has been attributed to the measure related to organisation of efficient environmental monitoring (71.9% of the interviewed citizens marked this measure as a high priority), which is being seen as a controlling instrument, not only for "polluter pays" principle implementation, but also for local government institutions. The second in a line is a measure that focuses on enabling the citizens to participate in environmental decision-making, which is actually a criticism of the previous, long-term practice in environmental policy planning. According to the participants, the third in a line of priorities is establishing of local environmental strategy. Such participants' opinion is formed based on their statement that practicing environmental protection is more talking than doing and their fear that such strategy is only going to be just another piece of paper, without actual realisation.

Environmental information and education (by its scope, quality, dissemination speed and availability) is a policy measure with rising significance, whereas it is the key figure for forming and raising environmental awareness. Participatory environmental assessment showed that, within the public, awareness of the role of this measure is just being formed. In the conversation with the interviewers, participants pointed out the significance of environmental education in schools.

In addition to policy measures that represent concrete actions in environmental protection, the second key element in environmental policy enforcing is the stakeholder's network. Its characteristics and the manner of functioning are determined, *inter alia,* by social identification of environmental risk. The community that considers some activity potentially environmentally hazardous for its well-being will arrange and organise social actions in order to prevent such hazard as well as delegate functions and responsibilities to the social stakeholders who might bear consequences of potentially environmentally risky activities.

Table 2. Citizens' opinion on environmental stakeholders' involvement

Measure	Extent of involvement				
	None	Little	High	Not responded	Total
1. Local (municipal) authorities	0.8	16.7	79.2	3.3	100.0
2. Public (municipal) enterprises	0.6	10.7	87.2	1.4	100.0
3. Citizens and citizens' organisations	1.5	33.4	63.5	1.6	100.0
4. Enterprises	3.0	27.5	64.3	5.2	100.0
5. Educational and research institutions	1.2	28.1	66.9	3.8	100.0
6. Provincial Secretariat for environmental protection	1.3	13.3	81.9	3.5	100.0
7. Media	1.2	39.9	57.0	1.9	100.0
8. Political parties	21.0	35.9	36.3	6.8	100.0

Table 2, contains data with the results of participatory environmental assessment regarding the stakeholder's network. The participants were asked to evaluate the extent of involvement of each stakeholder. Regarding their opinion, the largest extent of involvement is assigned to public enterprises that represent local executive authorities and apply policy measures into practice. The second largest scope is given to the media, which indicates the raising public awareness on significance of quality environmental information. Quite promising is an evaluation of the extent of involvement of the inhabitants themselves that comes third in line. It shows the willingness of public to take part in environmental policy implementation, mostly through different civil organisations. Political parties have been marked as the stakeholder with the lowest expected extent of involvement. According to the participants in environmental assessment, although they are legitimate instruments of struggle for fulfilment of social interests and goals, political parties are also seen as very inappropriate because of the tendency to manipulate environmental issues and exploit them for other causes (for example, the increase of number of voters).

4. Conclusion

Participatory environmental assessment can be seen as an instrument for an identification of public interest in resolving environmental issues. Regarding local environmental policy measures, the assessment points to the valuable course lines in their implementation:

- More activities in the area of environmental information and education, such as establishing regular information service system; such system would be responsible for environmental data compilation and their circulation;
- More activities in the area of enabling public to participate in environmental decision-making.

In addition, participatory environmental assessment showed that some activities should be undertaken in order to improve characteristics of stakeholders' network. The improvement means intensification of stakeholders' co-operation, which is vital to successful environmental policy enforcement.

References

1. Convention on Access to Information, Public Participation in Decision-Making and Access to Justice in Environmental Matters, United Nations Economic Commission for Europe, Switzerland (1998).
2. P. Markovič, Guide to Implementing Local Environmental Action Programs in Central and Eastern Europe, Regional Centre for Central and Eastern Europe, Belgrade (2001).
3. Local Environmental Action Plan — Manual, Regional Centre for Central and Eastern Europe, Belgrade (2004).

CHAPTER 33

INVESTIGATION OF JOB SATISFACTION IN THE STATE OWNED SECTOR OF SERBIA

OLGA HADŽIĆ

Faculty of Sciences, University of Novi Sad, Novi Sad, Serbia

MILENA NEDELJKOVIĆ

Faculty of Law and Business Studies, Novi Sad, Serbia

Job satisfaction is the topic of wide interest to both people who work in organization and people who study them and it is the most frequently studied variable in organization behavior research. Job satisfaction can be considered as a global feeling about the job or as a related constellation of attitudes about various aspects or facets of job (salary, promotion, supervision, benefit, co-workers, communication). The assessment of job satisfaction is a common activity in many organisations where management feels that employees' well-being is important. For example, as some studies have found, job satisfaction plays a critical role in an employees' decision to be absent and many models of turnover place job satisfaction in the centre of a complex process that involves factors both inside and outside the organization. Improving job satisfaction not only raises the company's profits, but also facilitates the company's development. In this paper we investigate the impact of age, gender, job tenure, education and position on job satisfaction in organisations belonging to the state owned sector in Serbia at the beginning of the radical organisational transformation.

1. Introduction

Job satisfaction has been widely studied over the last four decades of organisational research.[12,16] The notion of job satisfaction can be considered as a global feeling about the job or as a related constellation of attitudes about various aspects or facets of job, such as salary, promotion, supervision, benefits, co-workers, organisational procedures, nature of work (autonomy, variety, task identity, advancement opportunities, working conditions, etc.) and communication.

A large number of studies have investigated the relationship between job satisfaction and various organisational variables (organisational commitment, performance, organisational level, leadership style, organisational culture and climate, etc.) and demographic characteristics of employees (age, gender, job tenure, education and position). In many organisations turnover remains one of the most troubling issues, especially in research and development organisations, and since the inverse relationship between job satisfaction and turnover has been established, one of the most important tasks of management in organisations is the assessment of job satisfaction of employees.

In this paper we shall present some results of our investigations of the relationship between age, gender, tenure, position and education with job satisfaction in a big organisation from the state owned sector in Serbia, at the beginning of the radical organisational transformation. Finding ways to promote job satisfaction among Serbian workers may enable managers, both domestic and foreign, to improve their firm's performance, without incurring substantial additional costs.

2. Bibliography Review

The majority of studies on the relationship of age and job satisfaction have found some association between employees' age and job satisfaction.[1,7,9] Some[6] suggest that job satisfaction is U-shaped in age, with higher levels of motivation among young workers but that this declines after the novelty of employment wears off. Satisfaction rises again in later life as workers become accustomed to their role.

Clark *et al.*[3] supported a U-shaped relationship between age and job satisfaction and proposed that employees early in their careers (between the age of 20 and 30) had both high expectations and limited understanding of what made certain jobs good or bad. For many younger workers the employment may represent their first job out of school and receiving salary may influence the level of job satisfaction. As a career proceeds, the awareness of other job situations may promote a negative view of one's existing job. Other studies report a linear positive relationship between age and job satisfaction,[10,11,14,15] since, and by their opinion, aging influences higher income and more responsible job, which increases job satisfaction.

The relationship between gender and job satisfaction has been extremely inconsistent across studies. Greenhause, Parasuraman and Wormley[5] found no significant gender differences in job satisfaction, even though the females in their study were less likely to have managerial /professional jobs and more likely to have low-paid clerical jobs than the males. There are some explanations of the equivalent job satisfaction of women to men despite the non-equivalent job conditions and salary. Brush *et al.*[2] suggested that women may differ in expectations, i.e. women expect less from work and so they are satisfied with less. This may have developed over generations in which women had to accept fewer promotion opportunities and lower salary even for the same job. An interesting explanation is given by Witt and Nye.[17] By their opinion there can be gender differences in perceptions of equity.

General studies typically show that the more educated have lower job satisfaction[4] since they have higher job expectations.

3. Research Questions and Assessment Instrument

In our investigations we used Paul Spector questionnaire (JSS),[16] which assesses nine facets of job satisfaction as well as overall job satisfaction. The facets of job satisfaction are:

1. salary,
2. promotion,
3. supervisor,

4. benefit,
5. contingent rewards
6. operating procedure,
7. co-workers,
8. nature of work,
9. communication.

The scale contains 36 items and uses a summated rating scale format, which is one of the most popular for job satisfaction scales. The responses to the JSS items are numbered from 1 to 6 (from 1-disagree very much to 6-agree very much). Some of the items are scored in a positive and some in a negative direction. There were also questions about demographic characteristics of employees from the sample and the most important facet of job satisfaction. Questionnaires were distributed to 250 individuals in the organisation and 218 of them answered to all of the questions. Participation was voluntary and anonymity was guaranteed. The sample consisted of 38% of females and 62% of males. Ages ranged from 20 to 65 years, tenure (length of time) ranged from less than 5 years to over 25 years and education ranged from the basic educational level to the university degree of education.

Due to a limited space provided we shall present only some of the obtained results of our investigation on job satisfaction.

4. Summary of the Findings

4.1. *Distribution of the Most Important Facets of Job Satisfaction per Age, Gender, Education, Tenure and Position Demographic Groups*

The studied age groups are: 1-less than 30; 2-more than 30 and less than 40; 3-more than 40 and less than 50; 4-more than 50 and less than 65. In all age groups the largest percentage of the employees considered salary, nature of work, co-workers and contingent reward as the most important facets of job satisfaction, but their distributions (percent of employees) are significantly different. For example, 53.13%

of the youngest employees considered salary as the most important facet of job satisfaction, but only 38.1% of the oldest employees shared their opinion. For 33.33% of the oldest employees the most important facet is nature of work, but for only 15.63% of the youngest employees it was the most important.

In both gender groups the largest percentage of employees considered salary as the most important facet of job satisfaction (47.41% of male and 44.58% of female). Nature of work is the most important facet for 32.53% of employees from the female group, but only for 19.26% of employees from the male group.

For 66.67% of employees from the first educational group (i.e. employees with the first degree of education — eight years of elementary education) salary is the most important facet of job satisfaction, but that was the case for only 37.5% of employees with the university degree of education. None of the employees from the first group considered nature of work as the most important facet, but 27.78% of employees from the group of the most educated employees considered it as the most important. Salary is also the most important facet of job satisfaction for 29.71% of employees from the managers group, but 49.72% of employees from the non-managers group considered salary as the most important facet.

It is important to acknowledge that at some point, age and job tenure are unavoidably linked, particularly in the higher tenure brands, where it is impossible to have young employees. Hence, increases in job tenure depend on increase of age, which implies similarity between the distributions of percents of employees when considering the most important facet of job satisfaction. For example, 56.41% of employees from the "less than 5 years job tenure" group considered salary as the most important facet of job satisfaction, but only 43.18% of employees from the "longer than 25 years of job tenure" group. For 31.82% of employees from the "longer than 25 years of job tenure" group nature of work if the most important facet, but it is so only for 20.51% of employees from the "less than 5 years job tenure" group. From the findings of this study we conclude that further research should be conducted into the impact of interaction of age and job tenure on job satisfaction.

It is interesting that the facet of job satisfaction–supervisor, in the organisation under investigation, is not as important as in the case of the organisations from Western countries. Only 0.55% of the employees from the non-managerial group and 2.7% of the employees from the managerial group considered the relationship with the supervisor as the most important facet of job satisfaction. Research on the relationship with supervisors (the so-called feedback environment) in organisations in the context other than the US is important, as recent theoretical and empirical work in the feedback environment suggests that feedback processes in organisations are highly dependent on cultural values and social systems.

4.2. *Statistical Results Obtained by ANOVA*

By ANOVA we found statistically significant difference in the level of overall job satisfaction between male and female employees and also between managers and non-managers. Similarly as it is the case in

Table 1. Overall job satisfaction and gender

Group	Count	Average	Standard deviation
Male	135	2.622	0.818
Female	83	2.850	0.734

Table 2. Facet of job satisfaction-organisational procedures and gender

Group	Count	Average	Standard deviation
Male	138	3.386	0.830
Female	85	3.607	0.726

Table 3. Overall job satisfaction and position

Group	Count	Average	Standard deviation
Managers	38	2.973	0.677
Non-managers	185	2.655	0.807

Table 4. Facet of job satisfaction-nature of work and position

Group	Count	Average	Standard deviation
Managers	38	4.322	0.770
Non-managers	185	3.864	0.934

Table 5. Facet of job satisfaction-supervisor and age

Group	Count	Average	Standard deviation
1	32	4.695	0.789
2	57	4.382	1.012
3	89	4.180	0.944
4	44	4.005	0.935

many Western organisations, the level of overall job satisfaction and satisfaction with facet-organisational procedures of female employees is higher than that of male employees. The level of overall job satisfaction and satisfaction with respect to the nature of work of managers are statistically significantly higher than that of non-managers. One of the main reasons for such a result may be that the power in the decision-making process of managers in the state owned organisations in Serbia is still on a very high level.

From Table 5, we can see that the increase in age implies decrease in the level of satisfaction with the facet-supervisor. By ANOVA we also found statistically significant differences between the following groups: 1 and 3; 1 and 4; 2 and 4. One of the explanations for such a result may be that the older employees have a feeling of a high level of job insecurity during the transition period, which causes a high level of communicational needs. Since the main source of communication about the planned changes has to be the supervisor, the expectations of older employees about the quality of communication with supervisors are on a higher level than the expectations of younger employees.

By ANOVA we found statistically significant differences between the following groups: 2 and 4 and 3 and 4 with respect to the level of satisfaction with the facet-benefits. The employees in the organisation

Table 6. Facet of job satisfaction-benefit and age

Group	Count	Average	Standard deviation
1	32	2.906	1.069
2	57	2.618	0.954
3	89	2.764	0.904
4	44	3.145	0.962

who are between 30 and 50 years old have statistically significant lower level of satisfaction with the facet-benefit, than employees from "more than 50 years old" group. In the last 15 years, during the period of political and economical instability in Serbia, the second and third groups of employees were not in the position to obtain any 'intangible' benefits. On the other hand, the experiences of the oldest group of employees were quite different. Namely, in the period before 1990s in the 20th century the incentives for the best employees (for example, the vacation at the Adriatic Sea during the summer period which was paid by the organisation) was a common form of 'intangible' benefit and many employees who are more than 50 years old experienced such a benefit. So, the intangible benefit may account for the higher overall job satisfaction in the whole work period.

5. Conclusion

The obtained results related to the most important facets of job satisfaction are similar to those in other East European countries in the transition. Contrary to the situation in Western organisations, in which employees consider the relationship with supervisors as one of the most important facets of job satisfaction, in the state owned organisation under investigation, as it is the case in other East European countries in the transition, the work was characterised, in a long period, by a high level of job security, fixed salaries and intangible benefits, which means that supervisors had little control over their subordinates' career and salary.

In terms of Hofstede's cultural dimensions, Serbian workers rank very high on the uncertainty avoidance dimension in contrast to the US workers. Organisations operating in a high uncertainty avoidance culture will use more formal rules, procedure and structures for providing feedback information than in a culture which can be characterised by low uncertainty avoidance.

Hence, in the organisation under research, the feedback information, especially informal feedback information, is less dominant in predicting work attitudes. In spite of the fact that the feelings of uncertainty and job insecurity among the employees in the organisation at the beginning of the radical organisational change is very high, it seems that employees did not considered supervisors as one of the most important sources of information about the future position of the organisation (including the number of the employees in the recent future).

By Hofstede, Serbian workers ranks very high on the power distance cultural dimension.[8] Organisations operating in a high power distance culture will convey feedback more frequently through a top-down feedback process, whereas organisations in low power distance cultures will convey feedback more, like in the US.

From recent research of Mojić[13] we conclude that there is a shift among Serbian workers in private organisations with respect to index of the power distance and uncertainty avoidance cultural dimensions (i.e. they are much lesser then it was the case in the period of Hofstede's research) and we can expect that, after the privatisation of the organisation, the facet of job satisfaction — supervisor will be more important as a predictor of job satisfaction for all employees.

The increasingly global business world necessitates the ongoing training of domestic supervisors in new managerial skills, especially in the field of organisational analysis and organisational communication. The future expatriate manager has to be acquainted with the culture of the host community in order to behave appropriately. He must learn to communicate in the new cultural environment, since culture contributes significantly to explaining key differences in societal behaviour including organisational behaviour.

References

1. D. Bernal, D. Snyder and M. McDaniel, *Hum. Resour. Abstr.* **34**, 398 (1999).

2. D.H. Brush, M.K. Moch and A. Pooyan, *J. Occup. Behav.* **8**, 139 (1987).
3. A. Clark, A. Oswald and P. Warr, *J. Occup. Organ. Psych.* **69**, 57 (1966).
4. A. Clark and A. Oswald, *J. Public Econ.* **61**, 359.
5. J.H. Greenhaus, S. Parasuraman and W.M. Wormley, *Acad. Manage. J.* **33**, 64 (1990).
6. F. Herzberg, B. Mausner, R. Peterson and D. Capwell, Job attitudes: Review of research and opinion, Psychological Service (1957).
7. W. Hochwarter, G. Ferrie, P. Perrewe, L.A. Witt and Ch. Kiewitz, *J. Appl. Soc. Psychol.* 1223 (2001).
8. G.H. Hofstede, *Cultures Consequences: Comparing Values, Behaviors, Institutions and Organisations Across Nation* (Sage Publications, 2001).
9. G. P. Latham, *Work Motivation, History, Theory, Research and Practice* (Sage Publications, 2007).
10. R. Lee and E. R. Wilber, *Hum. Relat.* **38**, 781 (1985).
11. S. Linz, *Int. J. Manpower.* **24**, 626 (2003).
12. E.A. Locke, *Organ. Behav. Hum. Perf.* **4**, 309 (1969).
13. D. Mojić, *Sociology, XLV.* **4**, 317 (2003).
14. M. Nedeljković, Investigation of job and communication satisfaction in organisations, Master thesis, University of Novi Sad (2006).
15. G.E. O'Brien and P. Dowling, *Aust. Psychol.* **16**, 49 (1981).
16. P. Spector, *Job Satisfaction: Application, Assessment, Cause and Consequences* (Sage Publication, 1997).
17. L.A. Witt and L.G. Nye, *J. Appl. Psychol.* 77, 910 (1992).

CHAPTER 34

TRIPLE HELIX — THE FIRST PUBLIC PRIVATE PARTNERSHIP IN SERBIA EXAMPLE OF THE "BEST TECHNOLOGY INNOVATION COMPETITION"

VIKTORIJA BOJOVIĆ

Faculty of Economics, University of Novi Sad, Novi Sad, Serbia

Recent changes visible in the increased co-operation between the public and private sector in realisation of complex projects are discussed in the paper. Partnership, bridging the gap and eliminating time lag of application of basic research into industry, is crucial for a long-term economic growth. Triple Helix, as public private technology partnership, is an economic wealth generator. Socio-economic environment in the Republic of Serbia is fruitful for implementation of new university-industry-government partnership. The former is proven with an example of "The Best Technology Innovation Competition" the success of which is measured by the number of newly founded high-tech companies.

1. Introduction

Societies and economies are changing and there is a need to bring together the paramount elements and practices of that change. While the assumption that the public sector is responsible for delivery of basic services remains deeply rooted in the minds of many policy makers, the methods by which these services are created, obtained and delivered are changing. This is reflected in the desire to engage private sector to work in complex and demanding projects in the public interest.

Governments, worldwide, have gained an experience in the increased involvement of the private sector in procurement of public services. The underlying logic for establishment of partnership is that both sectors have specific positively correlated characteristics that could jointly facilitate more effective provision of public services. The most successful partnership agreements draw on the strengths of both public and private sector to establish complementary relations. Rather than considering partnership as a model, it should be thought of as a process.

Public private partnerships were identified as an alternative to traditional public procurement. The term "public private partnership" (PPP) has come into frequent use only recently, e.g. in the UK in 1990s. Prior to that, in an attempt to improve the delivery of public services, there has been a great deal of discussion determining the degree of involvement of private bodies in delivery of public services.

Defining the term "partnership" one could say in short that it is a joint venture with shared risks and profits. In this way co-operation, inclusion and association are related to partnership. The concept was encapsulated in the following definition[3]:

> "Partnering involves two or more organisations working together to improve performance through mutual objectives, devising a way of resolving disputes and committing to continuous improvement, measuring progress and sharing gains".

The focal advantage of public private partnership is threefold. Firstly, the partnership allows the government to maintain an active role in developing policy initiatives. Secondly, the partnership provides means for private sector to complement, rather than replace the government. Finally, it encourages a valuable exchange of skills and experiences between the two sectors.

2. Public Private Technology Partnerships

A new model of collaboration between the sectors can be realised in an array of different industries and fields. Most common are partnerships in demanding and complex infrastructure projects. The DBFO (design-build-finance-operate) model is considered to be the synonym for public private partnership in infrastructure, as it is the most suitable for complex projects and gains the most benefits.[1] Moreover, partnerships are suitable for policy making in most developed economies, such as the United States of America. The third field of application is technology transfer.

The importance of technology partnerships is in balancing activities of basic research at universities and colleges on one side and applied research in industry on the other. Traditionally, universities have been viewed as the source of new knowledge that is published openly and drives the creation of new technologies. Although there are limitations to measuring the contribution of basic science to economic growth, multiple studies demonstrate a positive rate of return for economy on the whole.

Partnership is crucial for a long-term growth as a tool for:

1. Bridging the gap between research and development at universities and in industry;
2. Eliminating a time lag of application of basic research; and
3. Giving incentives to the researcher.

Furthermore, it is a key ingredient in more applied research and development (R&D). Investments in research and development are critical for economic growth. Investments are made by industry and government at the same time with a different purpose. Commercial products and innovations are rarely a stimulus for basic research and a reason why firms have little incentive to fund basic research on their own. That is why the government funds majority of basic research through universities and colleges.

Partnerships in the applied technology research generally take two forms. The first involves the government funding of industrial research and development. The second involves government and industry scientists actually working together. The idea behind this type of partnership is not necessarily to meet the government objectives, but to help private sector move research from the lab to the market.

2.1. *Triple Helix as a Model of Public Private Technology Partnerships*

Triple Helix model is based on interaction between academia, industry and the government. Therefore, the interplay of universities, of the whole economy and of the government institutions can be thought of as a chain, or better as three helices that advance together through a dense and non-linear communication and negotiations. The model looks upon the United States national system of innovation and thus mostly popular in the American literature. European equivalent is Mode 2[a] explained by several principles[2]:

1. Knowledge produced in the context of application.
2. Trans-disciplinarity.
3. Heterogeneity and organisational diversity.
4. Social accountability and reflexivity.
5. Quality control.

Thus, it is crucial to embed the university-based scientific research in a larger social context, to link universities with business (industry) and to tie science and technology closer together. Ultimately, the boundaries between science and technology become blurred.

The difference between terms such as: science, technology, research, development and innovation can be presented as in Figure 1. As shown, science and technology (S&T) can be addressed as a subset of innovation while research and development as a subset of science and technology. Therefore, knowledge can be measured across different dimensions and dimension-assigned indicators: research and development, science and technology and innovation. In this paper we chose to measure knowledge with the innovation potential, embracing both research and development and science and technology potential.

In the triple helix model, communicating intensively, the knowledge producers (universities) and knowledge users (industry) are directly connected during the whole process of knowledge creation.

---

[a] As a supplement to Mode 1 which refers to production of scientific knowledge in a traditional University setting.

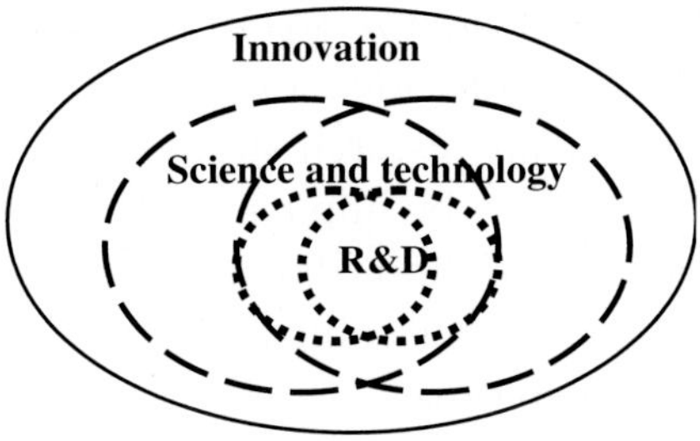

Figure 1. Illustration of different knowledge dimensions.

In other words: the interest to use knowledge effectively often demands participation in the process of knowledge production already.

Firms optimise their research and development processes to reduce "time-to-market" by paralleling their research and development activities. They use networks to access the necessary and highly specified knowledge from universities, but also from other firms. Second, codified as well as tacit knowledge defines important key resources for firms. The "paralleling" of basic research, applied research and experimental research can be considered as a two-fold strategy: on one hand supporting basic research, and, at the same time, accelerating the market introduction of research and development on the other hand. Paralleling processes can take place within the context of individual institutions or across different institutions and sectors (e.g., university/business interfaces). Network benefits for universities include taking over some corporate culture "values" and developing them in favour of an "academic entrepreneur".

3. Model Implemented in Serbia

The capacity to recognise and assimilate relevant knowledge from universities more specifically could be stimulated through Science and Technology Parks, or Technology Transfer Centres, or on more ground level through Business Competition in the High-Tech sector. For more on these topics see.[5]

First attempts to "produce" more academic entrepreneurs were made in 2003, at the University of Novi Sad. More than 80 future entrepreneurs were attracted to the regional "Best Idea Competition". The supporting idea was to change the prevailing perception that the prosperous research should be left at the university and wait for "better days" to be implemented. The goal was proliferation of entrepreneurial spirit and explanation of all the advantages of running a high-tech company compared to doing research at the University Solely for a Diploma.

The socio-economic environment in the Republic of Serbia was helpful. Being a transition economy several changes were performed in parallel. First, new laws that were entrepreneur and investment friendly were passed. Second, all socially owned enterprises were privatised by national or international entities. Significant inflow of foreign capital enhanced competition but left more than enough space for inexperienced small businesses. Practice and education on real problems was made possible for all daring enough. Economy in transition is very lucrative to invest into having more growth potential than old capitalist economies that already reached their peaks and are saturated with the intense competition. All this made favourable climate for founding new firms. Adding to everything previously said, the policy makers have focused on university based start-ups as a potential catalyst for regional economic development.

3.1. *The Best Technology Innovation Competition*

The implementation of Triple Helix model in Serbia was made through "The Best Technology Innovation Competition". As a form of partnership between the government (represented by the Ministry of Science and Environmental Protection), the University of Novi Sad and industry, the competition is traditionally held at a national level. The objective of the project is the promotion of high-tech entrepreneurship, evaluation of entrepreneurship potential and transmission of techniques needed for business planning. Special attention is paid to protection of intellectual property rights of innovations applied for the competition.

The objective of the government, as a partner in this venture, is to identify the winning projects that are socially beneficial because of high spillovers, but would not be realised solely due to the lack of entrepreneurial spirit. The prize fund is the main responsibility of the Ministry.

The university is the key player with its educational responsibility. It is not a hotbed of entrepreneurial energy and the faculty is not inherently adequately equipped for managing small venture portfolios. In order to continue making progress with commercialisation of the results of academic discoveries, the study of the process of commercialisation proves to be practically important. Specific trainings were custom made for high-tech and innovative start-ups. Unique trainings cover business area, such as market analysis for novel products/services/technology/software, segmentation, positioning, creating vision and mission, defining business goals, organisation of innovative firms, financial planning etc.

The competition is organised every year from March till December in two stages. The first stage covers gathering of all innovations and evaluating their innovative potential. The second stage offers entrepreneurial education free off charge to all potential innovators. Finals are organised at the end of December before the jury of Ministries (Minister of Science, Minister of Economy, etc.), representatives of financial institutions (CEOs of major banks and funds) and industry granting its contestants free propaganda nationwide.

The aspiration of counting a constantly greater number of created spin offs from the faculty year in year out is very hard to fulfil. Tracking all the start-ups as a direct consequence of competition is almost impossible. Determining thresholds in measuring success of competition is difficult as well. One might say that one company per year as a direct outcome would be a reason enough to organise the competition next year, because the whole economy is better off. Some screening has been done and is shown in Table 1.

The analysis of provided data suggests a high success of the competition. The number of applied innovations is satisfyingly high every year. Close to three hundred innovations per year implies great innovative potential of the Serbian economy. What is needed is to intensify different stimuli for market realisation. During the competition (from March till December) almost a dozen innovations have been

Table 1. Number of start-up firms as a direct consequence of the Competition

Year of competition	Number of ideas/teams	Number of contestants	Number of ideas/teams entered second round	Number of start-ups during the competition
2005	346	900	170	9
2006	254	900	180	8
2007	274	928	178	unknown yet

fully realised. That number of start-ups in high-tech industry could be regarded as sufficient contribution to the small transitional economy. Many more companies and firms are founded, but this is hard to track. The achieved impact on attitude evolution is something more important. Spreading the idea of self realisation of innovative ideas among scientists and university workers is the key to overall success of the competition, proving its need on a yearly basis.

The university contributes more than just funding, often providing infrastructure, business plan assistance, support staff, and other resources to enable the research faculty to start firms. These efforts coincide with the increasing pressure on modern public universities to promote measure and report on their contributions to the economic development in their region.[b] For more on this topic see Ref. 4.

4. Conclusion

Changing socio-political conditions in the Serbian economy forced policy makers to implement a new concept of private sector involvement in procurement of public services. Public private partnerships are predominantly intended for complex and demanding projects.

[b] The chief concern is that encouragement and resources provided to the faculty to start firms may result in faculty spending less time on university research, leading to fewer research outputs, research outputs with marginal contributions to the field, or both. Some faculties have openly criticised university technology transfer as a negative influence on academia.

One field of application can be technology transfer between universities and industry. A widely spread public private technology partnerships can take form of Triple Helix. The collaboration between the government, university and industry was realised in the Republic of Serbia in a form of "The Best Technology Innovation Competition". Having the sole intention of proliferating entrepreneurial spirit, its success could be measured in the number of newly founded high-tech companies. Organising the competition nationwide in three consecutive years enriched the transitional economy of Serbia in quantitative terms as well as in quality achieving impact on the attitude evolution.

References

1. D. Campbell and W. Guttel, "Knowledge production of firms: Research Networks and the "scientification" of business R&D", *Int. J. Technol. Manage.* **31**, 152 (2005).
2. J.E. Stiglic and S.J. Wallsten, Public — private technology partnerships — promises and pitfalls, Public Policy Partnerships, The MIT Press (2002).
3. M. Geddes, *Making Public Private Partnerships Work: Building Relationships and Understanding Cultures* (Grower Publishing Company, 2005).
4. P.C. van der Sijde and A. Ridder, *Infrastructures for University Spin-off Programs: An Introduction* (NIKOS, 2002).
5. R. Lowe and S. Gonzalez-Brambila, "Faculty Enterpreneurs and Research Productivity," *Int. J. Technol. Manage.* **32**, 173 (2007).
6. Shane Scott "*Academic Entrepreneurship: University Spnoffs And Wealth Creation (New Horizons in Entrepreneurship Series)*", Edward Elgar, USA (2004).
7. V. Bojovic, "Production Private Partnerships as a Last Resort for Traditional Public Procurement", *Panoeconomicus*, No. 3, (2006), pp. 299–311.
8. V. Senk, M. Despotovic, B. Ristic, "Science & Technology Park Novi Sad: Current Status and Long-Term Goals", *Science Policy and Human Resources Development in South-Eastern Europe in the Context of European Integration*, (2006), pp. 157–162.

CHAPTER 35

WHAT LIES BENEATH: MALE AND FEMALE ADDRESSES IN SERBIAN

BILJANA RADIĆ-BOJANIĆ AND NATAŠA KARANFILOVIĆ

Faculty of Philosophy, University of Novi Sad, Novi Sad, Serbia

The paper focuses on the role of male/female addresses to the same and opposite sex and investigates how these addresses reflect the Balkans culture. The research is based on a questionnaire filled in by 100 students in their third and fourth years. The preliminary analysis shows that there seems to be solidarity among the same sexes, while in the opposite-sex addresses there is no solidarity. When men address women, they appeal to their sexuality, whereas when women address men, they primarily appeal to their intelligence, or the lack of it. To conclude, it seems that both praises and insults are based on cultural gender stereotypes.

1. Introduction

This paper explores various forms of addresses that men and women use when talking to and about each other. These forms of address reveal cultural background and social structuring of a community and offer insight into gender roles which define interpersonal geography. Of special interest for this paper is "the nature of the linguistic forms used by both genders to refer to each other and establish social patterns of perception and address capable of showing the interplay between linguistic and social structures."[1] Holmes[2] claims that "gender is a complex continuum which interacts with other social dimensions such as social status, ethnicity, age and power," all of which expose certain aspects of the culture which is investigated. In the same vein, this paper will attempt to link the linguistic forms used by the informants with the background data on the culture in which the forms are used and with the purposes of use.

Each language tends to have highly idiosyncratic and complex sets of address forms which are inevitably used when men and women talk to each other, either in same-sex or opposite-sex dyads. Gender, then, surfaces as a strictly cultural category, revealing one of many aspects of the culture under scrutiny. The focus of this paper, therefore, lies at the intersection of several fields of research: linguistics proper, sociolinguistics and cultural anthropology. Linguistics focuses on the forms and scope of addresses, sociolinguistics on the contexts and reasons for the use of addresses, and cultural anthropology on the link between addresses and the cultural context in the background. Holmes[2] explains that, as opposed to conversational analysis of male/female exchanges, "a more satisfactory way of studying the linguistic realization of gender thus involves examining the way individuals express or construct their gender identities in specific interactions in particular social contexts." In other words, the spotlight here falls on how language is used to build the gendered world that individuals construct for themselves and for others and the best way to approach this issue is through sociolinguistic research. Just like in this paper, sociolinguistic investigation aims to show how linguistic differentiation reflects social structure and, consequently, the cultural patterns as seen in linguistic and extralinguistic behaviour.

Cameron[3] describes a "cultural difference model," which explains the different lenses through which men and women perceive one another as if they were from different cultures. Evidence from research on different cultures suggests that sex differentiation through linguistic means is almost a pan-cultural recourse.[1] It is, therefore, of interest to see how male and female addresses in Serbian reflect individual male and female cultures as well as the shared culture in Serbia. Forms which men and women use to refer to each other are lenses which alter and redefine the view of the referent but are also a means of disclosing how self-perception and perception of the addressor function.

The paper is based on a research conducted among 100 third and fourth year students at the Faculty of Philosophy, University of

Novi Sad. They filled in a questionnaire which contained the following questions: (1) what is the form you use to address someone; (2) which sex do you use it for (man, woman or both); (3) how frequently do you use it (frequently, sometimes, rarely, never); (4) which type of situation do you use it in (positive, negative); (5) which structures do you use it in; (6) which situation do you use it in (students describe the context). The questionnaires were then analysed and the results were placed in the theoretical frame presented above. Finally, it should be said that the paper does not intend to defend of refute models of male and female linguistic behaviour, but merely to describe the state of affairs in the Serbian language and culture today.

2. Analysis

The research for this paper was conducted among approximately 100 third and fourth year students of the English language and literature, who are mainly young people in their early twenties. The decision for this sample of population lies in the fact that older generations seem to show more respect for their addressees, at least on the surface, and tend not to use insulting forms of addresses. The questionnaire was conducted anonymously, where students received one piece of paper with instructions and another with a blank table which they filled in with the relevant information. Each question in the questionnaire had a certain purpose: the first question was supposed to elicit the address form itself, the second one aimed to determine the sex of the addressee, the third one established the frequency of usage, the fourth one described whether the situation of usage was positive or negative, the fifth question defined the structures and the sixth one the context in which the listed forms of addresses were used. For the purposes of this paper, and because of the limitations of space and time, the discussion of the structures will be omitted, but all other categories will be taken into consideration when drawing conclusions on the way men and women address each other and refer to each other.

The first level of analysis reveals the fact that forms of addresses and referring can be broadly divided into 6 categories: women addressing and referring to men, women addressing and referring to both sexes, women addressing and referring to other women, men addressing and referring to other men, men addressing and referring to both sexes, men addressing and referring to women. The analysis will show if there are any overlaps, similarities or parallels between male and female addresses. Of priceless help is the last category in the questionnaire, which offers information regarding the context of usage and describes not just experiences of informants, but their hidden motivations and cultural agendas.

When men are addressing or referring to women, two large groups of meanings can be discerned: positive and negative. Within each of these there are several subgroups which bear a lot of significance in disclosing cultural relationships and attitudes. Firstly, when men talk to or about women in positive situations, they primarily appeal to their appearance, in most cases attractive and appealing. In such contexts men use expressions like *riba* [chick],[a] *cica* [pussy cat], *mačka* [cat], *mačkica* [kitty], *primerak* [specimen], *lutka* [doll] or *pička* [pussy]. Although men consider most of these terms to be highly complimentary for women they are addressing or talking about, a closer look reveals that most of the terms are actually metaphors derived from the terms for female genitalia, which narrows the role and existence of the woman to a mere sex organ. The dominant and praised characteristic of the woman in the eyes of the man is, thus, her sexuality reflected in her looks and expressed in a constricted lexical set. Another much less frequent meaning found in this group is the seeming implication of solidarity among men and the women they address or refer to. In this groups there are lexemes such as *cura* [girl], *mala* [little one], *devojčica* [little girl], which reveal that the solidarity male informants name as the reason for using such terms is just seeming. What underlies these forms of addresses is the male protective attitude towards the women they are talking to or about. That can hardly qualify as solidarity since there is no equality — male

[a] The translations given in the brackets are, in most cases, cultural approximations of the term in Serbian, but sometimes there were no available corresponding terms in English, when the author resorted to the literal translation which, to a certain degree, reveals the nature and the meaning of the term in Serbian.

informants are assigning themselves the superordinate role of the protector who looks after the weak and unprotected woman.

In negative situations men are usually motivated by three different factors in female behaviour or appearance: evil, stupidity and sexually explicit actions. The first motive is the least frequent one; here male informants listed three terms for women who are acting in a bad way and who are being mean: *kuja* [bitch], *džukela* [dog], *aždaja* [dragon]. In all three cases there is mention of animals that, in folk belief, have a tendency for angry and aggressive behaviour, which is then reflected onto women and their verbal behaviour. The second motive, stupidity, again provokes the use of a range of animal names (*krava* [cow], *koza* [goat], *kobila* [mare], *krmača* [sow], *guska* [goose], *kokoška* [hen]), all of which again rely on folk beliefs that such animals are stupid.[4] Interestingly, female counterparts in the pairs of animals are used in each case. The third motive and the most frequent one is the sexual behaviour of the woman who, in the eyes of the male informants, acts like a prostitute, whether implicitly or explicitly. Hence the range of terms: *kurva* [whore], *kurvetina* [hooker], *kurava* [slut], *kučka* [bitch], *drolja* [tramp], *fuksa* [hooker], *fufa* [floozy], all of which are quite self-explanatory.

In cases when male informants are addressing or referring to both sexes, they normally use collective nouns such as *stoka* [cattle] or *masa* [mass], without an individual identity or individual traits.

When male informants are talking to or about other males, they are using three large groups of terms with the following general meanings: (1) admiration (men looking up to other men), (2) solidarity and equality, and (3) disparagement (men looking down on other men). The first groups of address forms reveals that admiration is based primarily on personality traits (*car* [emperor], *kralj* [king], *baja* [hot shot], *lik* [character], *momčina* [big guy], *faca* [dude]) and that men are here perceived by other men as having heroic qualities, being very deserving and as having achieved great things. Admiration is never based on other men's good looks or their verbosity. In case of solidarity and equality, male informants mostly use the following terms: *brat* [brother], *matori* [old man], *drug* [friend], *ortak* [mate].

Like with admiration, equality and solidarity bear the meaning of closeness regarding the social position of men included in the exchange, of long acquaintanceships and of relationships created through business, a primarily male activity.

The third group of terms denoting disparagement shows how men look down on other men and what lack of appreciation is based on. The most dominant meaning that underlies the largest number of expressions relates to the lack of intelligence in male addressees: *kreten* [cretin], *idiot* [idiot], *moron* [moron], *debil* [imbecile], *duduk* [stupid], *budala* [fool]. In continuation, another large negative group of terms refers to the annoying behaviour that male addressees display, thus triggering the use of terms such as *drkadžija* [jerk-off], *smrad* [stinker], *krme* [pig], *smor* [bore], *davež* [pain in the neck], and *džiber* [stinkpot]. The questionnaires show two interesting phenomena regarding male/male addresses. Firstly, the only sexually motivated address is *paćenik* [poor sod], which denotes a man who suffers because he does not have a sexual life. Secondly, there are two terms which are repeatedly used to address cowardly and treacherous men (*kurva* [whore], *pička* [cunt]). Interestingly, these two terms are also mentioned as being used often to talk to and about women, when they refer to women's sexuality and promiscuity.

When women are addressing or referring to men in positive situations, they are primarily dealing with their physical appearance, especially if they are talking to or about a handsome, good-looking, well-dressed man. In those situations women resort to a whole series of expressions, ranging from *dasa* [hunk], *frajer* [hunk], *faca* [hunk], *tip* [guy], *tipčina* [big guy] to *macan* [kitty cat], *slatkiš* [cutie] and *beba* [baby]. The first group of expressions shows that women view men as very masculine, which they find attractive, while the other group of terms shows a motherly, protective attitude towards men, who are seen as little, unprotected creatures. In neither situation are men seen as exclusively sexual objects, which contrast with the male view of women, who are in the same situation most frequently referred to with various terms for female genitalia. Additional positive situations rely on men's other positive qualities, such as their friendliness (*drugar* [mate], *matori* [old man]), success (*car* [emperor]), kind-heartedness (*ljudina* [big guy]) and creativity (*ludak* [lunatic]). All of these expressions are also found in the group of expressions when men are addressing or referring to other men in positive situations, which is probably the original lexical set that women borrowed these expressions and meanings from.

As for negative situations in which women talk to or about men, the most dominant one is men's bad or annoying behaviour, in which case they are qualified as *kreten* [cretin], *smrad* [stinker], *gad* [scab], *debil* [moron], *drkadžija* [jerk-off], *konj* [horse], *majmun* [monkey], *somina* [catfish] and *gnjavator* [bore]. This group of expressions resembles the expressions men use in the same situations. Additionally, men are described by women as *životinja* [animal] and *skot* [bastard] if they are behaving badly or aggressively towards a woman, which is the situation not found in the questionnaires filled in by male informants. The only references to men's sexual behaviour are contained in the terms *jarac* [he-goat], who is a lewd, older man interested in younger women, and *pačenik* [poor sod], who is a sexually frustrated man with no sex life. Interestingly, men's promiscuity is something that is not mentioned once in a negative light. There were, however, several instances of positive contexts in which terms for men's promiscuity were used, where men were labelled as *jebač* [fucker], which probably integrates the man's sexual success and his good looks.

The positive situations when women are addressing or referring to other women are in most likelihood a transfer from the male system of values, so women are again positively assessed as sexy, good-looking and attractive. It is somewhat surprising that women are using expressions like *cica* [kitty], *riba* [chick] or *mačka* [kitty cat] when talking to or about other women. An additional motive for the use of various forms of addresses lies, as expected, in the solidarity among women, when expressions like *matora* [old woman], *carica* [empress], *sestra* [sister] and *drugarica* [friend] are used. These terms either show the admiration on the side of the addressor towards the addressee, who is a woman worthy of praise, or solidarity and closeness towards someone who is the same as the addressee or in a similar situation.

In negative situations women attack other women for several reasons: stupidity, bad behaviour, promiscuity, clumsiness and unappealing appearance. Stupidity provokes the use of a range of animal names,[4] such as *guska* [goose], *kokoška* [hen] or *ćurka* [turkey], probably because of the folk belief that these animals are stupid. Additionally, women are referred to as *glupača* [stupid woman] or *gluperda* [really stupid woman] in such situations, which are transparent and self-explanatory terms. When women display bad behaviour, be it evil, frustrating or verbally annoying actions, they are called *kučka* [bitch], *pizda* [cunt], *pička* [cunt], *aždaja* [dragon], *rospija* [shrew], *gadura* [virago] or *lisica* [vixen]. In cases when women are promiscuous, they are referred to with a whole range of names, some of them being *fuksa* [floozy], *drolja* [tramp], *prljavuša* [dirty woman], *kurveštija* [whore], *radodajka* [easy] and *prostakuša* [slut]. It is interesting to notice that in case of promiscuity women are not referred to with the means of female genitalia; such addresses are used when female and male characters are treacherous and deceitful. This can probably be connected with the belief that a man can easily be swayed and influenced through sex and that women use sex to manipulate men. Women who are observed as clumsy or unattractive are usually termed *krava* [cow], again because of the folk belief that cows are clumsy and unattractive animals. A final set of terms, found occasionally in the corpus, consists of *jadnica* [pathetic woman] and *pačenica* [poor sod], which refer to a woman who is sexually frustrated.

The terms women use to address both sexes can be divided into positive and negative ones. In positive situations the main motive is affection so women use *mačak* [cat], *mali* [little one] and *slatkiš* [cutie] to address the people they really like and feel close to. In case of negative emotions, both sexes are referred to with general terms, not particularly connected with either sex, similar to men. Here terms such as *budala* [fool], *stoka* [cattle], *đubre* [trash] or *smrad* [stinker] are used, especially if the addressee has irritable or boring behaviour.

3. Conclusion

As Cameron[3] says, "systems of address reflect the salient social distinctions of the culture (kin versus non-kin, intimate versus non-intimate, higher/lower status)," as is in the case of this research. The system of addresses here discloses views that men and women have of the same and opposite sex and, even more, what kind of view it is in each case. Gender roles are defined through the interaction of informants' age and the power each sex has in the given culture. Identities are here constructed from two different stands: the male and the female. On one hand, women are frequently treated as intimates or subordinates, which are seen in the use of a number of terms motivated by women's sexuality or their seeming need for protection. These two situations are perceived as a good opportunity for establishing the relationship of intimacy and men seem to think that it is their given right to do so. Such forms of address describe the woman as being in a subordinated position, either because she is weak or because she is deduced to a mere sexual object. Cameron[3,5] notes that "more words are available to insult women than men, especially in sexual terms, and that words for women's bodies are more taboo than those for men's."

On the other hand, men are perceived in a completely different light: in positive situations they are seen as attractive and worthy of admiration and, in fewer cases, as equals with whom one establishes the relationship of solidarity because they are strong, independent protectors. They do not invoke a promiscuous image and in just one case of such address this kind of behaviour is not just tolerated but wholeheartedly saluted. This linguistic phenomenon, Cameron[3] states, is connected with the sexual double standard: "[w]omen should have no sexual desires whereas men should be insatiable; if women get out of line by expressing autonomous desire or refusing to cater to men's, they will be censured." The whole cultural background is, therefore, full of contradictions which are the basis of sexist stereotypes and the view of women as mere sexual objects or the "weaker sex." This image is adopted by young women as well who alter the self-perception under the influence of men's reality expressed in language.

It seems as if this very small portion of language reveals a great deal about the way men and women perceive the same and the opposite sex. It also discloses cultural facts and stereotypes in the background which, at the beginning of the 21st century seem to still hold their ground. It might be worth seeing what addresses will sound like in 10 years' time and if the perceptions will show a tendency for change or stagnation.

References

1. Z. Maalej, *Tunisian Review of Modern Languages* **9** 155 (1999).
2. J. Holmes, *Introduction to Sociolinguistics* (Longman, 1992).
3. D. Cameron, *Rethinking language and gender studies: Some issues for the 1990s, Language and Gender: Interdisciplinary Perspectives* (Longman, 1995).
4. S. Halupka and B. Radić, *J. Pragmat.* **35** (1891).
5. D. Cameron, *Feminism and Linguistic Theory* (Macmillan, 1995).

CHAPTER 36

CANNIBALISED BODIES AND IDENTITIES, MARGARET ATWOOD'S, *THE EDIBLE WOMAN*, *LADY ORACLE*, AND *CAT'S EYE*

ANDREEA ŞERBAN

Faculty of Letters, History and Theology, University of Timisoara, Timisoara, Romania

Women have constantly been perceived as the social "other" and defined against men's superiority. Margaret Atwood uses food imagery and eating to suggest power relations in a society which seem to obey the laws of the natural world (i.e. "eat or be eaten") rather than the laws of humans. Focusing on three of Atwood's novels — The Edible Woman, Cat's Eye, and Lady Oracle — the paper deals with metaphors of cannibalism and the role they play in the female protagonists' reconsideration of their identities. The three novels selected are representative for three kinds of symbolic cannibalism in the sense of amputated or frozen selves. Each of the three protagonists will learn her lesson of assertiveness through (symbolic) trips to the origins, and become stronger women able to control their own lives.

1. Introduction

Cannibalism may essentially be described as an uncontrollable hunger, similar to vampirism, which sets a boundary between prey and predator, *us* and *them*. Cannibalism has been one of the essential features used by Europeans in their presentations of the colonised — the Other(s) — who were frequently described as child-like or animal-like and presumed to eat the flesh of their own kind, i.e. other humans.[19]

Images of cannibalism have been used to demonise those who are different culturally, racially and even sexually.

If we are to consider cannibalism as a semiotic tool in establishing relations and apply the "us — them" criterion to contemporary society, then the social Other would be the woman. Although fascinating, women have always been considered inferior to men one way or another and defined in relation to men, first the father and then the husband. Biological determinists have claimed that biology grounds and justifies the social norms, i.e. men's superiority over women; in other words, social differences between men and women stem from physiological differences. Thus men pursue the new, while women preserve the old; men are energetic, active and independent, while women are conservative, passive and dominated by emotions.[10]

Men and women have been attributed specific gender roles by the patriarchal society, where women may have homemaker roles such as housewives or mothers, where they do not (or cannot) conduct important business or make important decisions, where they need men for everything, i.e. protection, money, advice, or where they are considered sex objects or decorative ones with no personalities ("women who know their place", the sexist discourse would claim). It is through their own bodies that women are made vulnerable, partly because of their dependency on pregnancy and partly because the male culture fetishes such parts of women's bodies as breasts, legs and bottoms.[4,9] Thus men assess and rank women in an attempt at controlling their bodies and at eliminating the threat (of castration) they might pose. On the other hand, women have been "trained" to feel attractive only if they are watched by men, i.e. if they attract and retain men's attention; the more men look/gaze at them, the more attractive they feel. Women thus exist to be consumed either by way of gazing, or by way of symbolic eating within love relationships.

2. Atwood's Preys and Predators

Cannibalism, as well as the related myth of vampirism, relies on becoming stronger by literally or metaphorically consuming others. If one's body is cannibalised/vampirised, i.e. fragmented, the power and energy of an individual are also fragmented. In Margaret Atwood's novels, cannibalism — as well as vampirism — prevails in its symbolic meanings and plays an important role in terms of power relations between the characters. All Atwood's protagonists are initially fragmented, victimised, cannibalised, but they gradually transform into stronger individuals through eating — becoming predators — or metaphorical touching (writing/narrating or painting). In other words, edible victims become inedible consumers.

Atwood's novels depict a society in which the laws of nature are extremely important in terms of power relations, namely "eat or be eaten", where the prey is a social other. I have selected in this respect three of Atwood's novels to illustrate three types of metaphorical cannibalism which I have called hetero-cannibalism, homo-cannibalism (as opposed to hetero-cannibalism) and self-cannibalism respectively. Thus, in *The Edible Woman* men are the cannibals, fragmenting women's bodies by imposing stereotypes of femininity on them. In the second case, *Lady Oracle*, it is women who are the cause of metaphorical dismemberment or amputation of other women's bodies (for example, the daughter's body) by perpetrating the same stereotypes of femininity. Finally, in addition to being constantly criticised in terms of appearance and behaviour (i.e. emotionally cannibalised) by her so-called friends, the female protagonist of *Cat's Eye* acts as cannibal against herself in her trying to cope with a harsh reality.

Women's cannibalisation or, in other words, the amputation of their identities is primarily suggested through images of mouth and teeth and speech implicitly. Marian, the protagonist of *The Edible Woman*, seems mouthless in the sense that we never hear her agree to Peter's marriage proposal. Moreover, she has a rather feeble voice since the second and largest part of her story is narrated in the third person singular (as opposed to the first and third part of the novel which are in the first person), suggesting the distance Marian takes both from herself and from society. Elaine Risley (*Cat's Eye*) becomes the victim of the girl-friends she has longed for when she travelled through the wilderness with her brother and parents and, even though they torment her and continually try to regulate her appearance and behaviour by constantly criticising her, she cannot bring herself to talk about her victimisation. She thus resorts to self-cannibalism, using her teeth to tear off the skin from her heels, to chew her hair or nibble at her fingernails in an attempt to preserve her integrity and sanity, which seem to be dissolving.

Similarly, *Lady Oracle*'s protagonist, Joan Foster, cannot bring herself to talk about her multiple identities (fat unpopular child, lover, poetess, popular gothic writer, assumed terrorist) to her husband. She only informs him about her poetry volume when the book is under print and only in the end of the novel does she take the decision of letting him know about her writing gothic romances. Throughout the novel, her motivating force seems to be the fear of being found out, of people discovering the truth about her fat childhood self, about her writing both poetry and popular gothics.[9] But unlike Elaine, Joan falls victim to homo-cannibalism, with her mother (embodying society, the "others") trying to inflict society's ideals of femininity on her daughter and control her body through eating, through what goes into Joan's mouth. Desperate for her mother's love and unable to obtain her approval, Joan begins to eat out of panic and her body swells like dough. Eating becomes not only a form of self-mutilation, a paradoxical self — cannibalism and erasure of one's identity through flesh, but also a valuable weapon against her unloving mother.

As a child, Joan is often invited to watch her mother put on make-up, a ritual she secretly enjoys. The mother's double mouth, which she creates with red lipstick, as well as her red fingernails, vividly remind us of cannibals' blood — dripping mouths or blood — thirsty vampires who have recently fed themselves on human victims to acquire their powers and assert their own control. Looking forward to escaping her authoritarian mother, Joan gives in to society's indoctrinating models of femininity and, stimulated by Aunt Lou's inheritance, she loses weight much to her mother's spite, who feels her control diminishing. Even though Joan has rejected beauty stereotypes as a fat child and teenager, she has internalized her mother's control and ideology, and will remain their victim throughout her adult life as a slim woman.

In order to reconsider her situation, the mature and slender Joan stages her death and leaves for Europe. With the mock suicide, she almost succeeds in completely erasing at least one of her multiple identities (that of Joan Foster), but will have to return to Canada to save her "accomplice" friends from prison. In Italy, after she dances on shattered glass, a symbol of her split selves, and gets her feet minced, Joan decides to try to reconcile her multiple identities and tell her husband about her writings so that she may live without fear of being discovered.

Just like adult Elaine of *Cat's Eye*, whose speech is controlled by her lover Josef, who tells her when to speak and when to keep quiet, trying in this way to assert his power and control over her, Marian MacAlpin from *The Edible Woman* succumbs to Peter's authority and does not have a say even in what concerns her appearance. After the engagement, Peter flaunts Marian in front of his friends and colleagues and wants her to change her hairdo and buy more feminine clothes instead of the greyish ones she usually wears as a sort of camouflage to pass unnoticed.

Eating and teeth denote a voracious appetite and Atwood uses such images to overlap physical and sexual appetites. Thus Arthur (*Lady Oracle*) grinds his teeth during sexual intercourse, while Peter (*The Edible Woman*) has "teeth-gritting moods", bites Marian's shoulder during sex and finds her perfume appetising. He also enjoys kissing her on the neck, a very vulnerable place, which reminds us again of blood-thirsty vampires and which transforms Peter into a blood-sucking creature, highlighting his need to feel in control (especially of Marian's body).

Atwood draws a parallel between courting and hunting with a focus on the predatory aspect of both rituals. In the restaurant scene before the marriage proposal, Peter describes his hunting and gutting of a she-rabbit. Marian identifies with the rabbit, subconsciously feeling trapped and in danger of being split in halves. This feeling is even more acute as she has dreamt about her body dissolving, "turning transparent" and vanishing into thin air as an indication of her own identity being amputated by Peter's authority and wishes. When Marian offers Peter a heart-shaped cake as a belated Valentine gift, he devours it voraciously. If we consider the pink cake a symbol of Marian's own heart, then Peter is nothing but a hetero-cannibal who absorbs his fiancée's vital energy and gains more and more power over her. While Marian's body begins to have a will of its own and reject food, Peter has no such problems and metaphorically devours the heart-shaped cake in a possible attempt to incorporate Marian and achieve the perfect union,[7] or "the mythical state of oneness,[14] that will enable him to assert his will and better control her.

Atwood's protagonists' search for selfhood is usually accompanied by a search for food or by eating itself. As Parker points out, the female characters are hungry throughout the novel and they cannot find anything nutritious to eat.[12] Thus Marian's body rejects food, Joan takes her decisions at the table while eating junk food while Elaine abuses her body with a poor diet. However, it is through food association and eating that the protagonist of *The Edible Woman* also takes her life into her own hands. Finally acknowledging her victim position, Marian rejects it and bakes a woman — shaped cake to serve her fiancé. The cake stands for artificial womanhood, a substitute victim for Marian. When Peter refuses to eat it, she incorporates it herself in an act of self-cannibalism, regaining control over her body and asserting a new identity as a stronger woman who has rejected the stereotypes of femininity that society in general and Peter in particular have sought to impose on her. In an analogous way, Joan Foster's attempt to take control and assume responsibility over her life is accompanied by eating: "I sat at the white table with my hot cup, adding another white ring to the varnish, eating a package of rusks and trying to organise my life".[2]

Atwood's protagonists learn the lesson of assertiveness through consumption. No longer consenting to their cannibalisation, they turn from prey to predators. Nevertheless, apart from the powerful images of eating, mouth and teeth discussed above, another way of amputating or dismembering women's identities is through images of eyes and seeing.

As sensory organs, the eyes are the ones that establish a direct contact with the viewer and with the world in general. Through the eyes, a person witnesses and learns about the world. As the title suggests, *Cat's Eye* is centred on the eye with its primary function of perception and looking, but also as a protective amulet and suggestive of artistic vision. The novel is a first person narration and, therefore, the reader sees everything through the mediating eyes of the main character, Elaine Risley, who returns to her home-town Toronto partly to re-see her past and (re)discover her true identity. In so doing, she begins her search for Cordelia, her childhood friend and tormentor.

Coming from an idyllic wilderness, as the family accompanied the father who was a biologist, Elaine has trouble adapting to urban life and feels like a social outcast. She cannot get rid of the "heathen" label given to her by Mrs. Smeath, Grace's mother, and this will torture her throughout her adolescence and adult life. In her desperation to become accepted, she must follow the rituals of the society to which she wishes to belong, i.e. she must learn to play the games girls her age play.[5] Elaine is thus put under a "magnifying glass" and constantly criticised — emotionally cannibalised — by her three "friends": Grace, Carol and Cordelia, who seem to remind us of the three "weird sisters",[16] or of the Greek Moirae/Fates[1] who "controlled the metaphorical thread of life" from one's birth until his/her death. They "delete" Elaine's body by burying her in one of their games. This symbolic death and its subsequent rebirth should entail the character's assertion of a stronger and more powerful identity, an initiate, but what happens to Atwood's character seems to be quite the opposite. After she manages to get out of the hole, Elaine becomes subservient to Cordelia — "the main witch" — who transforms Elaine into her slave and absorbs her vital energies, almost completely erasing her individuality. Potvin argues that this metaphor of zombification is a model for "the internalising of oppression for women, for the creation of a pathological "slave-mind" within the ideological framework of patriarchy".[13] This idea is shared by Wilson who claims that the hole in the ground becomes symbolic for "the void of identity and women's powerlessness in patriarchy".[20] Nevertheless, Elaine is not the only the "slave" in the novel. Although she appears tough and demanding, Cordelia too is in fact enslaved by patriarchal demands; she is always careful to behave and dress properly, and she is afraid of her father who punishes her with beating. In fact, the girls only apply the rules of patriarchy and of Mrs Smeath, who has developed a patriarchal mentality and consequently enforces its rules and norms. Deeply hurt because she is seen as "wrong" and "sinful", Elaine attributes Medusa-like eyes to Mrs. Smeath who thus becomes a powerful "witch" that can petrify people and that would haunt Elaine throughout her adult life.[18]

In her need to be accepted and loved, Elaine is numbed and unconscious and undergoes self-mutilation by developing cannibalistic instincts; she begins to bite her lips and tear the skin around her nails and off her heels. It is through the feet that humans connect to the Earth — Mother, and having swollen or wounded feet suggests the identification with the dark, instinctual or animal side.[17] We can therefore argue that in her trying to get the approval of others, Elaine denies her connection with the Great Goddess and rejects precisely her inner femininity, trying to become stronger by dominating her weaker side. The time when she experiences the first splits between her body and soul is indicative of her split identity.

Denying the vision of Virgin Mary that she has after falling into the freezing cold water of the creek, Elaine chooses to survive as a dead living creature or a zombie, as she will later depict herself in her self-portraits. Similarly to a vampire, Elaine begins to absorb all the energies of her tormentors and, having now power, she grows indifferent to them. She begins to wear vampiric black and relishes in scaring Cordelia with a story of her vampiric other. In psychology, black is commonly linked to feelings of guilt and fear. That is why Elaine's wearing black clothes may suggest her (unconscious) fear of not being accepted by others and her guilt of not doing her best to make herself accepted. Elaine has exchanged her role of prey for the predator; she has now taken the place of her tormentor and has turned into a vampire, drawing in Cordelia's vital energies and making her grow weaker and weaker.

As I have argued in a previous article,[15] Elaine's trying to remember her past is paralleled by her attempt to find Cordelia, her alter-ego, her mirror image, which can make her whole again. The protagonist describes herself and Cordelia as two halves of the same face, the real body and its reflection in the mirror, the body and the soul. Since Elaine cannot find Cordelia during her stay in Toronto, our "vision" of her remains therefore only partial because, although we have learnt to see through her eyes, her image is not reflected, it is not complete, and we are left with an amputated identity.

The eye serves as a powerful symbol in expressing power relations. Men seem to fragment the protagonist's body; both her first lover and her husband want to change Elaine by turning her into a "body of desire".[6] It is not what women are that matters, but what they inspire in men.[11] Josef the art teacher sees women as "helpless flowers, or shapes to be arranged and contemplated".[3] Besides controlling Elaine's speech as we have seen before, his touch freezes her — as if in a painting — and then erases her. He first puts her into a perfume advert and then tries to "rearrange" her into a doll-like, pre-Raphaelite woman. The hotel mirror offers Elaine an insight into men's perception of her, but this image is not a clear one as the mirror is smoked, symbolically suggesting an image imposed on the protagonist by others. Her thinking of the poppy she should be holding, as a pre-Raphaelite symbol, would then stand for her resurrection as a new woman totally complying with the norms of patriarchy.

When she becomes an artist, the protective cat's eye becomes Elaine's signature and represents her skill, in other words, artistic vision. It even replaces her head in a painting, suggesting not only the freezing of emotions, but also Elaine's attempt at asserting her identity through art. When, after her father's death, Elaine descends into the basement of her parents' house — as if into the Underworld to ask for the wisdom of the dead — and finds her lost cat's eye marble in an old purse, she sees her life entire in it. Having re-found her roots, symbolized by the basement of the paternal house, Elaine has re-membered herself and won back her lost self by removing the glass, i.e. the cat's eye marble, out of her eye. The moment hints at the opening of the third eye, or the passage from "the blindness of consciousness to the insight of imaginative seeing"[8] and also alludes to the (re)incorporation of what has been lost. The protagonist can now look at her own paintings from a different perspective: wiser and more objective, as she learns to forgive.

If at the beginning, all of Atwood's protagonists are edible prey, lacking such body parts as mouths (and implicitly speech), feet or eyes (and implicitly sight), they finally transform themselves and by rejecting their victim positions, they re-assert their identities as stronger individuals — they become predators, i.e. inedible consumers — with a voice and a vision of their own.

References

1. Wikipedia the Free Encyclopedia, http://en.wikipedia.org/wiki/Moirae.htm (2006).
2. M. Atwood, *Lady Oracle* (Seal Books, 1977).
3. M. Atwood, *Cat's eye* (Seal Books, 1989).
4. S. Bordo, Twilight Zones: The hidden life of cultural images from Plato to O.J., *Never Just Pictures* 107 (University of California Press, 1997).
5. R. Culpeper, Recogntion and rejection of victimization in the novels of Margaret Atwood http://my.tbaytel.net/culpeper/RecognitionAndRejectionOfVictimizationInTheNovelsOfMargaretAtwood.html (1989).
6. L.J. Davis, Visualising the Disabled Body: The classical nude and the fragmented torso, *Seminar on Body Studies*, University of Debrecen (2002).
7. S. Freud, Three: Contributions to the theory of sex, Modern Library (1938).
8. C.A. Howells, Margaret Atwood, second edition, Palgrave Macmillan (2005).
9. S. Lee Bartky, Body politics, *Seminar on Body Studies*, University of Debrecen (2002).
10. T. Moi, Sex, Gender and the Body. *The Student Edition of What is a Woman?* (Oxford University Press, 2005).
11. L. Mulvey, Visual pleasure and narrative cinema, http://www.panix.com/˜squigle/vcs/mulvey-vpnc.html (1975).
12. E. Parker, *Twentieth Cent. Lit.* **41**, 349 (1995).
13. L. Potvin, *J. Pop. Cult.* **36** (2003).
14. S. Sceats, *Food, Consumption, and the Body in Contemporary Women's Fiction* (Cambridge University Press, 2000).
15. A. Şerban, The Magic of the Eye. Margaret Atwood's Cat's Eye, Universităţii de Vest, 77 (2007).

16. W. Shakespeare and Macbeth, *The Complete Works of William Shakespeare* 858 (Wordsworth Edition, 1999).
17. A. de Souzenelle, Simbolismul corpului uman, Amarcord (1999).
18. H. Staels, Margaret Atwood's Novels, Tübingen (1995).
19. J. Valsiner, *Cult. Psychol.* **6**, 88 (2000).
20. S.R. Wilson, Margaret Atwood's Fairy-Tale Sexual Politics, University Press of Mississippi, Jackson (2005).

CHAPTER 37

FEMININE AND ARTISTIC IDENTITIES: TRACY CHEVALIER'S NOVELS

DANA PERCEC

Faculty of Letters, History and Theology, University of Timisoara, Timisoara Romania

The paper discusses three novels by Tracy Chevalier — Girl with the Pearl Earring, Falling Angels and The Lady and the Unicorn — from a double perspective: the feminine condition and the role that art plays in the lives of characters. Women fight to obtain or preserve their freedom and to overcome loneliness in a similar manner in three different historical periods of the West European history. The novels construct an imaginary genesis for three works of art: Vermeer's famous portrait of a "Girl with a Pearl Earring", the anonymous French tapestry and the presence of some statues in an English graveyard at Queen Victoria's death.

1. Introduction

Tracy Chevalier, the American born writer whose novel *Girl with the Pearl Earring*[1] was unanimously considered by critics to be a revelation, has proved to manifest a special interest in two themes of great impact in a time of blurred aesthetic boundaries and of a sometimes desperate search for one's "real" identity — the status of the female individual and the role of art in social life. The three novels to be discussed — *Girl with the Pearl Earring*, *Falling Angels*[2] and *The Lady and the Unicorn*[3] — approach the two themes in three different historical periods and locations — 17th century Holland, Victorian England and late medieval France.

2. Feminine Identities

The feminine identity in all three books is built or preserved through suffering, pain and sacrifice. From Griet, the servant girl (*Girl with the Pearl Earring*) who has to marry a man she doesn't love only to save her family from disgrace to Kitty Coleman (*Falling Angels*) who joins the suffragettes' movement in order to make a statement in front of a traditionalistic husband and an authoritarian mother-in-law, women in Tracy Chevalier's novels obtain a voice at a very high price. Griet, in love with her master, the Dutch painter Vermeer, accepts to be his model for a painting although she knows this would compromise her and possibly even prevent her from ever getting married. She also accepts to wear her mistress's pearls in her ears for Vermeer's painting although she knows she will be fired if the lady finds out and although her ears are not pierced. She performs the operation herself, risking an infection and enduring great pain in front of the artist who seems to be completely unaware of or insensitive to her sacrifice. Kitty, a young wife in a well-off English family, will pay with her life for choosing another path than that designed by patriarchy — mother of many children with exclusively domestic preoccupations. Although she joins the suffragettes hoping she is doing the right thing for her daughter who will be able to vote and go to university, thus being potentially free from the authority of any man, Kitty is very little understood even by Maude, her only child. During a march organised by the suffragettes, a horse will run her over and Kitty will die the same evening, not knowing if she has managed to get her message through to her daughter or if she has been completely misunderstood by all the members of her family.

In all the three novels, women suffer severely from loneliness. In *Girl with the Pearl Earring*, Tracy Chevalier is inspired by Vermeer's inclination to paint lonely women engaged in various activities that go from physical chores such as pouring milk or sewing to artistic or intellectual activities such as playing the guitar, reading and writing. The loneliest of all, however, seems to be the mysterious young woman in *Girl with the Pearl Earring*, the portrait of a female character who is not engaged in any activity and who is painted against a dark grey background, only her face and left shoulder being visible. For this portrait, the writer invents a story of unrequited love (an 18-year-old girl who falls in love with a master much older than her and whose only concern is the chromatic perfection of his creations). Griet has to leave her parent's house and become a maid in a Catholic house because, after her father's work accident, the family has lost the only source of income. The girl feels her loneliness very acutely when she leaves her neighbourhood, heading for the Papist's Corner, where she has to spend the nights in a basement room being terrified by a crucifix, an unusual sight for a young person with an exclusively Protestant education, or when she has to accept Pieter the son's courtship only because her family wants an alliance with a well-off butcher who is the living promise that they will always have meat on their table (which has not happened since the father's accident). The loneliest moment is, however, her mute suffering in Vermeer's studio when she has to pierce her ear and hide her pain, knowing, at the same time, she will have to leave the house and never see him again as, the painting once over, she will no longer be able to stay around.

In *Falling Angels*, Kitty's miserable loneliness is best signalled by her obsession with warmth. At a garden party, in the evening, with people sitting around a big fire, Kitty draws closer and closer to the source of warmth until she realizes everybody else is a couple of yards behind her. Being colder than others, she feels at a certain moment that, if she made one step forward, she would catch fire. She is unhappy because she can't find the right place: she is either too far from or too close to the source of heat. If she stayed with others, she would be cold. By the fire, she is both uncomfortably hot and isolated from her family and friends. She feels rejected even by her daughter whom she loves but does not understand: when Kitty takes her, once, by the hand, Maude seems to be unpleasantly surprised and runs away as soon as she can; when Kitty takes her to town to introduce her into the suffragettes' movement, she is aware that Maude accepts only because she has never travelled by automobile.

In *The Lady and the Unicorn*, the imaginary genesis of the famous French medieval tapestry, Genevieve de Nanterre, the rich and noble lady whose husband, Jean Le Viste, has ordered the piece of art, feels the more isolated the more numerous are the ladies in waiting and the servants around her. She has failed in the only thing the society and family expected from her – giving birth to a male heir — and therefore she is no longer needed by anybody. It is known that in pre-modern and early modern Europe, women's inability to have children bore connotations of failure and inadequacy; it was a stigma of shame.[4] Dame Genevieve's husband neglects her, never omitting to remind her of his disappointment. She has no friends and the only way in which she could make good use of her isolation, going to the convent and serving God, is denied to her: "It would be a mercy to let me enter a convent. But Jean is not a merciful man. And he still needs me. Even if he despises me, he wants me next to him when he dines at home, and when we entertain or go to Court to attend the King. It would not look right for the place next to him to be empty. Besides, they would laugh at him at Court — the man whose wife runs off to a nunnery. Most men would be like that — older women joining convents are usually widows, not wives. Only a few husbands will let them go, no matter their "sins". She sees that people despise her or are sorry for her being an inefficient wife. Her eldest daughter, Claude, a girl with a boy's name who should have been a boy, shares the others' opinion. When Genevieve gave birth to the third daughter, Claude, although only seven, manifested her dissatisfaction openly by sniffing, as she had seen adults do, at the sight of the baby whose sex did not make her father happy. This may be also because female firstborns in aristocratic families, who were supposed to inherit the fortune, adjusted sometimes to this situation and responsibility less easily even than fathers.[4]

The other feminine character whose loneliness is conspicuous in *The Lady and the Unicorn* is Alienor, the daughter of the Brussels weaver who works on Jean Le Viste's tapestry. Although surrounded by a loving family, a physical handicap isolates her from the rest of the world: she was born blind. As, given her handicap, she can't ensure her family a good alliance with another weaver, she feels useless and unhappy. She works twice as much as others, both in the kitchen and the garden during the day and in her father's workshop, where she has learned to sew although she can't see, during the night when others can't work without light and rest in their beds. However, when the opportunity of a profitable alliance appears, with a wood dyer who would supply blue paint for free (the basic colour in tapestries and clothing in those times, alongside with the red, replacing also the green which was not used) she prefers to avoid this horrible perspective (Jacques Le Boeuf is old, brutal and smells of the fermented sheep's piss he uses to fix the colour) by giving birth to an illegitimate child. Although a submissive daughter, she prefers this ultimate measure which implies both her own disgrace and financial ruining of her father's workshop.

For most of Tracy Chevalier's feminine characters, freedom has a very high price. Kitty becomes a suffragette because she can't approve of her mother-in-law's view on marriage: women don't need to vote because their husbands are perfectly capable of doing so on their behalf and because "in any sound marriage, the woman is in perfect agreement with her husband" (p. 214). Edith Coleman voices here a general political comment against women's voting in the 19th century, that is, the fact that all women benefit from the political representation of a man (father, husband, brother) and therefore don't need to represent themselves.[5] The feminist counter — argument is that, on one hand, men cannot possibly know exactly what women want because they are so different and, on the other hand, there are so many women abused by the men in their family that one is entitled to have serious doubts as to the degree to which these men are indeed the women's protectors.[5] According to Edith Coleman, voting, like reading, won't get a woman anywhere. It will only put ideas in her head, which will succeed in making her unhappy. And who needs an unhappy wife with ideas instead of one who resumes herself to run the household with a firm hand, watching on the maid's virtue and on the cook's economical use of butter and spices in food? In this context, according to Bock,[5] Queen Victoria's model, with nine births, her attempt to place her husband, Albert, on an equal position with hers, giving him some of the important prerogatives of the English monarch, is an antifeminist example. Her figure of a loving wife, mother and widow supports the myth of the angel of the house, a middle-class ideal for many men and women that placed the woman in the centre of the domestic stage and nowhere else. Immobile but active within four walls, the mistress of the house — the only prestigious role a woman could dream of in a traditional society — has to work for the promotion of the typical bourgeois image of elegance and conspicuous consumption.

In this sense, Gertrude Waterhouse, Kitty's poorer but more conservative, Victorian neighbour, who loved Queen Victoria very much and wore mourning after her death more than any of her acquaintances and whose husband's name is, not accidentally perhaps, Albert, would be the prefect example. She makes efforts to preserve a façade of opulence when her husband's income is much more modest than the Colemans': "I do like to make an effort with my At Homes. I always have them in the front parlour, and use the rose-pattern tea set, and Elisabeth bakes a cake — lemon this week. Albert asks sometimes if we oughtn't to use the front parlour as the dining room instead, rather than eating in the back parlour, which is a bit cramped when the table is pulled out. Now Albert is right in most things, but when it comes to running a household I do get my way. I always feel better having a "best" room to show visitors to, even if it's only used once or twice a week. Thus I have insisted that we leave the rooms as they are, though I admit it is a bit inconvenient to fold the table back three times a day" (p. 111).

At the suffragette's march, Kitty feels totally free for the first time in her life. She is aware that none of the members of her family approve of her behaviour — in terms of social conventions, a riotous environment is no place for a lady.[4] Revolutions are for working class women; the better off the husband, the less possible it is for a woman to express her dissatisfactions in public without her gesture being considered incongruous and disgraceful. Dressed like Robin Hood, Kitty leads a symbolic procession — a fighter against a wrong social and political order. In Shakespeare's *As You Like It*,[6] Rosalind also ran away from a tyrannical uncle and the restraints of the Court and adopts a Robin Hood like attitude and attire. In the legendary hero's clothes, Kitty feels liberated of the constraints of the etiquette and the corset: "What I did feel sharply was the sun and air on my legs. After a lifetime of heavy dresses, with their swathes of cloth wrapping my legs like bandages, it was an incredible sensation" (p. 301). Bandages have been removed and she can walk unhindered. After being injured or ill, she is healthy for the first time in her life. But this feeling lasts for only a moment – the horse's hoof hits her chest and no doctor can save her any longer. She dies wishing that at least her daughter may find, one day, and the ideal place near the fire, where it is neither too cold nor too hot.

To Genevieve de Nanterre, both alternatives for freedom are denied: she can't rest in the peace of the convent and she can't commit suicide by throwing herself into the Seine because, suspecting this, none of the five ladies in waiting leaves her unattended even a minute. She is a luxury prisoner whose gaolers can't hide their boredom at having to be around her: "I heard one of them just now, huffing behind me from boredom. For a moment I felt sorry for them, stuck with me. On the other hand, they have fine dresses and food and a good fire in the evening because they are with me. Their cakes have more sugar in them, and the cook is generous with the spices — the cinnamon and nutmeg and mace and ginger — because he is cooking for nobles". Being a gaoler saves these women from the every-day hardships of life in pre-modern Europe, where survival was the most important preoccupation for the great majority. The ladies in waiting, although not servants from the lowest class, can't be ladies of their own house because their social position is too low to secure them a substantial dowry, the only real guarantee of good matches that persisted as a custom in Western Europe until the late 19th century.[7] For Claude, Genevieve's daughter, freedom would be the romance with the Paris painter her father hired, a romance for which she is ready to sacrifice a marriage with a nobleman from the Court Jean Le Viste chose as his son-in-law. She has to spend almost a year in the convent so that she can't meet her beau. A reward for her mother, a frustrated, introverted mature woman, the convent becomes a dreadful prison for a hot-blooded young and beautiful daughter.

Running away from Vermeer's house where she knows she will be accused of theft although it was the painter who imposed that she should wear his wife's pearls, Griet stops in the main square of Delft, where a star-shaped sign indicates various directions. She was not free when she was sent by her parents to the painter's house. She was not free when she sat for him in his studio, having to pierce her ears with the risk of an infection because the painter considered that a pearl would make the picture brighter. But, for only one second, she is free now when there is nothing to lose: "I could go back to my parents. I could find Pieter at the Meat Hall and agree to marry him. I could go off on my own somewhere far away. I could go back to Papists' Corner. I could go into the New Church and pray to God for guidance. I stood in the circle, turning round and round as I thought. When I made my choice, the choice I knew I had to make, I set my feet carefully along the edge of the point and went the way it told me, walking steadily". Although she chooses to do something she does not want deep down, the fact that she made this decision herself and it was not others who imposed it on her gives her a sense of peace, relief and security, perhaps even some sort of satisfaction.

In Tracy Chevalier's novels, the world seems to revolve around the feminine characters. No matter how old, educated or beautiful they are, they all have a special vitality. Except for *Girl with the Pearl Earring* where the story is told exclusively from Griet's perspective, in two other novels, all characters are given a voice. This multiple subjective perspective is a manner in which the author approaches an ever enriched femininity. The polyphony of the novels reveals a vast community of women, so different from one another in spite of the apparent similarities. There are mistresses; loving mothers or frustrated mothers, there are disobedient daughters who end up by becoming submissive, daughters exclusively devoted to their mothers, and obedient daughters who end up disobeying their fathers. Taking into consideration the limited number of scripts allowed to women by the society of most historical periods engaged in the social construction of gender — domestic or disruptive[8] — Tracy Chevalier's gallery is more than impressive.

3. Artistic Identities

Tracy Chevalier is the author of exceptional imaginative works that make art and heart inextricably interwoven. The pretext of the stories is the same: it is always a work of art that triggers the unfolding of the narrative speculation, giving birth to a parallel universe of authentic atmosphere and fabricated facts.

While *The Lady and the Unicorn* retraces the story of a tapestry, *Falling Angels* centres upon the symbolical presence of sculptures in a graveyard at a time of mourning for the British nation (i.e. Queen Victoria's earthly departure). The debut of this series is marked by the acknowledged masterpiece of the *Girl with the Pearl Earring*, which seems to set both the tone and the technique for the writer's ulterior creative research. Tracy Chevalier makes her readers take part in a literary experiment: intrigued by Vermeer's painting, she uses the relatively little existing information to subjugate the audience with the imaginary and incurably melancholic genesis of a famous work of art, thus easily transcending both temporal and spatial dimensions.

Griet is gradually initiated in the game of colours and illusions that make up life, love and art, consequently forming an inseparable and highly mysterious triangle. As if the writer was studying the potential intensities of light, everything happens in between glamour and darkness, ultimate purity and sheer degradation, aspiration and humiliation, abundance and modesty. The novel unfolds like a painting in itself, just as *The Lady and the Unicorn* puts together the various pieces of the narrative puzzle until its very end — that of the completion of the story which coincides with the crucial moment of the "cutting-off": "Since we are always working on just a strip of tapestry the size of a hand's length, which is then rolled inside itself onto a wooden beam, we never see the tapestry whole until it is done. We also work on it from the back and don't see the finished side unless we slide a mirror underneath to check our work. Only when we cut the tapestry off the loom and lay it face-up on the floor do we get to see the whole work. Then we stand silent and look at what we have made".

The above-described procedure concentrates the basic ingredients of Chevalier's recipe: the artist and his work inseparably connected by feelings, looks and silences. In her ineffable writing, the supremacy of the senses is proclaimed even above the reign of the written word; every gesture, every moment in time is visualised and made to come alive by means of a real or craftily constructed “magic eye”. The symbolic presence of the *camera obscura* mediates the sublimation of real-life nature and situations into art; it offers both understanding and interpretation by means of redoubling reality with a particular type of sensitivity, one that perceives and glorifies differences. After facing the miracle of the strange box, Griet begins to untangle the quizzes of painting: “He saw things in a way that others did not, so that a city I had lived in all my life seemed a different place, so that a woman became beautiful with the light on her face”.

Inflamed by the touch of a genius and beauty, things and beings gain unexpected proportions; they appear to populate a universe of their own, completely set apart from the sordid dullness of narrow streets, noisy fish-markets and winding canals. Art seems to change the quality of the portrayed, lingering on in the minds and hearts of a charmed audience, just like the face of the woman in the painting Griet sees in her master’s studio. “After she had gone, her face lingered like perfume”, notices the quick-spirited daughter of a reputed but blind tile-painter. One can, thus, easily see the way in which the mere shade of a thought or the simple fraction of an act are accompanied by sharp stimuli aimed at the senses, in an everlasting attempt to impress them directly upon the reader.

This straightforward communication between art and its admirer is openly felt by Griet – the first painting she ever sees stirs her first reaction on the way to affection: “I wanted to wear the mantle and the pearls. I wanted to know the man who painted her like that. I thought of me looking at my reflection in the mirror earlier and I was ashamed”. Chevalier turns out to be the master of an impressively wide range of exchanges of looks: tightly connected to the overwhelming realm of the visual, they function like implicit footnotes, like parenthetical comments and explanations to a series of ongoing events. The glance or the indirect look, the quick eye or the avoidance to meet a certain gaze are means of expressing feelings which are too intense or too outrageous to be spoken out loud. The novel stands out as a territory of silences in which the poundings of the heart can become unbearable at times, without ever fully bursting out into explicit words.

Griet and Vermeer are the protagonists of one of the most delicate love-stories ever written. Similarly to the case of Nicolas des Innocents, Claude Le Viste and Alienor de la Chapelle, those who are involved in affairs of the heart are brought together by the works of art they produce, supervise or benefit from. Thus, art is used as some sort of a superior matchmaking device, stirring passions which it will later on render, in specific shades, in its final product. There is, therefore, a constant interplay between feelings and the way in which they shine through in the work of art: in Chevalier's stories, the vicious circle is perfect, since one cannot clearly tell whether symbolism arises from life or whether it is life that modifies its course after being touched by the magic of art.

The writer is a magician of suggestion just as Vermeer is a wizard of image. He uses "false colours" in order to eventually give his perfected composition the proper tonality and discretion: "When at last he began to add colours on top of the false colours, I saw what he meant. He painted a light blue over the girl's skirt, and it became a blue through which bits of black could be seen, darker in the shadow of the table, lighter closer to the window. To the wall areas he added yellow ochre, through which some of the grey showed. It became a bright but not a white wall. When the light shone on the wall, I discovered, it was not white, but many colours". This is precisely the method Tracy Chevalier applies to the portrayal of her characters: their depths are hardly tackled with, they remain anonymous but, nevertheless, one can feel them boiling under the level surface of everyday life, shaping it, disturbing it, filling its contours — in a genuine but never aggressive way. Griet herself resembles one of the clouds given as an example of skilful achievement: pure, wise, docile, loyal, she willingly or unwillingly hides those features that create her unmistakable halo. Under the downplaying appearance of a maid lies the secret, indispensable woman, whose mythical hair, once unveiled, unleashes her unsuspected sensuality, unfortunately channelled in the end.

In between allusions and silences, the book breathes a never-ending sadness. It is the despair of contrasts, the painful result of an open confrontation between the illusionary realms of art and the disillusioned ones, buried in a reality which teaches a young girl about meanness, hypocrisy, cruelty, jealousy all too soon. It is a world of rough lessons, one in which survival has very little to do with any kind of effusion of the soul: it is a closed universe where beauty blesses the lives of the fortunate. Yet, fortune involves money, not the sensitivity or the inner inclination necessary for the appreciation of art. However noble and perfect the design, the owners and models used by the painter are nothing but abusers and adulterers, frightened servants, naïve maids or foolishly conceited aristocrats. This is not where superiority resides and Chevalier misses no opportunity to point it out.

An interesting case of genuine knowledge is illustrated, paradoxically enough, by two blind characters, Griet's father and Alienor. They appear to be doomed by a tragic destiny, lacking precisely the sense that makes the world of the novels go round: that of sight. However, they compensate by means of a sharpening of all the other virtues: it is her father's illness that helps Griet get her job with the famous painter, since she knows the importance of light and of the exact place of the objects in a studio. As for Alienor, she is a kind goddess of patience and good-will, conquering the reader by means of modesty inconsistent with her innate qualities. These are characters that mediate the connection of the visual with both the terrestrial and the spiritual: they live through *feeling*, both in the concrete sense of touching and in the abstract one, of emotional involvement. It is this literalness of feeling that distances Griet, as warm and beautiful a human being as she is, from the man that she has grown to love and utterly support.

Waiting, suspense, frustration form a background voice, a tragic chorus to accompany the lead singers' performance. The extraordinary is but slightly suggested, identities take shape within the carefully preserved boundaries of predictability, in between what is and what could be, if only the text, visual in itself, grows along with the delicate touches of brush that make up the mystery of human existence. Grievances are not shouted out, disappearances are not mourned, love is not declared and passions are not consumed. Even when these limitations are overstepped, what is left behind are the scars of separation and loss, since transitory feelings stand hopeless in front of the eternity of art. There remains a particular awareness of arriver-plans, a predisposition for melancholic interpretation of surfaces: "After that I could not stop looking at things".

References

1. T. Chevalier, *Girl with the Pearl Earring* (Harper Collins Publishers, 1999).
2. T. Chevalier, *Falling Angels* (Harper Collins Publishers, 2001).
3. T. Chevalier, *The Lady and the Unicorn* (Harper Collins Publishers, 2003).
4. O. Hufton, *The Prospect Before Her. A History of Women in Western Europe* (Fontana Press, 1997).
5. G. Bock, Femeia în istoria Europei din Evul mediu până în zilele noastre Polirom, Iasi, 178, 122 (2002).
6. W. Shakespeare, *The Complete Works* (Wordworth Editions, 1996).
7. J. Goody, Familia europeană. O încercare de antropologie istorică, Polirom, Iasi, 111 (2003).
8. S. Mendelson and P. Crawford, *Women in Early Modern England* (Clarendon Oxford, 1998).

